Introductory
Digital Electronics

List of Contributors

The contributors listed below are members of
The Introductory Electronics Course Team

G. Martin (*Chairman*)
J. Anderson (*Course Manager*)

Authors

D. I. Crecraft
N. Heap
L. Ibbotson (*NE London Polytechnic*)
A. Q. Jones
R. Loxton
M. L. Meade
C. A. Pinches
A. Reddish
G. Smol

Other members

G. Bellis (*Scientific Officer*)
J. N. Helsby (*Scientific Officer*)
R. Hoyle (*Designer*)
R. R. McShane
J. Newbury
A. A. Reilly (*Editor*)
E. Smith (*BBC*)
Professor J. J. Sparkes
J. Stratford (*BBC*)

N. W. HEAP AND G. W. MARTIN

Introductory
Digital Electronics

From Truth Tables to
Microprocessors

The Open University Press
Milton Keynes

The Open University Press
a division of
Open University Educational Enterprises Ltd
12 Cofferidge Close
Stony Stratford
Milton Keynes MK11 1BX
England

First published 1982
Copyright © 1982 The Open University Press

British Library Cataloguing in Publication Data

Heap, Nick
  Introductory digital electronics.
  1. Digital electronics
  I. Title   II. Martyn, Glyn
  612.3815   TK7868.D5

ISBN 0-335-10184-4

Text design by W. A. P.

Typeset by Oxford Verbatim Limited
Printed and bound in Great Britain at The Pitman Press, Bath.

Contents

Preface   vii

Acknowledgements   viii

Chapter 1   **Introduction**   1

Chapter 2   **Combinational logic**   4

Binary Inputs and Outputs	16
Truth Tables; Boolean Notation	23
Electronic Combinational Logic	34
Propagation Delay	46
Summary	48
Problems for Chapter 2	50
Appendix to Chapter 2	52

Chapter 3   **Sequential Logic**   56

Memory Devices: Flip–flops and Registers	59
General Properties of Sequential Circuits:	
Counters and Sequencers	64
Design of Counters	74
Summary	84
Problems for Chapter 3	85

Chapter 4   **Analogue/Digital Conversion**   87

Digital to Analogue Conversion	88
Analogue to Digital Conversion	101
Sample-and-hold Devices	110
Multiplexers	113
A Complete Analogue-to-digital Interface System	116
Summary	119
Problems for Chapter 4	120

Chapter 5 **Digital Components and Systems**   121

    A Simple Memory   123
    Random-access Memory (RAM)   132
    Read-only Memory (ROM)   146
    Introducing the Microprocessor   149
    A Microprocessor System Configuration   154
    Problems for Chapter 5   174

Appendix A   176

Appendix B   178

Index   183

Preface

This book is an edited version of part of the teaching text used for the Open University's undergraduate course 'T283 Introductory Electronics', first presented in 1980. The original text was produced by a course team of nine authors and nine support staff. The team was also responsible for student experimental kits, television and radio programmes.

The approach adopted by the course team was to try and teach, where possible, through specification of the problem rather than through discussion of the operation of a selection of available devices and components; since this leads more naturally to modern design strategies such as 'top-down'.

The emphasis in the book on the solution of combinational and sequential logic problems by the truth tables and ROMs, rather than logic gates and mapping techniques, illustrates this approach.

The book covers topics ranging from logic to microprocessor memory systems and is intended for students with a background in analogue electronics who wish to update their knowledge to include digital electronic systems. Chapter 2 introduces the basic ideas of combinational logic design; truth tables, ROMs, logic gates and Boolean algebra. Chapter 3 deals with sequential logic, and shows how one can design binary and decimal counters and use these to produce a system controller. Chapter 4 examines the system elements needed to interconnect analogue and digital systems. Methods of converting one signal type to another are described, as well as the techniques of sampling and multiplexing. Chapter 5 is a brief 'look over the fence' at the recent developments in 'microprocessors'. It is not intended to teach how to build or design a microcomputer system, but examines some of the essential elements.

Acknowledgements

The editors wish to acknowledge the helpful comments given by their colleagues from the original course team; to Mr. R. Loxton the acting Head of Discipline during the period the book was in preparation; and to Professor J. J. Sparkes, Dean of the Faculty of Technology at the Open University. We would also like to express our appreciation to Chris Martindale, Jane Barden and Diane Holl for their skilful services in typing the manuscript.

CHAPTER 1

Introduction

Electronic products and techniques are making a tremendous impact on all our lives in the latter half of the twentieth century. It is a period of rapid and exciting change for all of us involved in electronics, and as such it sets the authors a difficult challenge. This is to communicate to you in this short text our enthusiasm for electronics and its developments along with the basic technical knowledge required.

We will be analysing and designing electronic systems in terms of the system elements and their characteristics. Many of these system elements can be bought 'off the shelf', and we will not explain why they behave in the way they do, but will concentrate on how to use them.

System and Circuit Diagrams

Two of the key activities of electronics engineers are the design and analysis of complex electronic systems. Both these tasks are simplified by the use of system and circuit diagrams, since they provide a pictorial means of describing or 'specifying' the technical aspects of the system.

System Diagrams

A simple example of a system diagram is shown in the margin, that of a torch. It conveys the technical information that a torch consists of a battery, a switch and a bulb. Because the elements in the system diagram are often drawn as simple rectangular blocks or boxes, the diagram is sometimes referred to as a *block diagram*. The actual

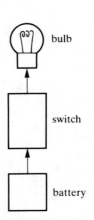

bulb

switch

battery

1

contents of the boxes are largely unimportant compared to their function, so the blocks are referred to as *black boxes*. In this book system diagrams, blocks and black boxes will be used as design and analysis aids.

There are a few important points to remember about such system diagrams;

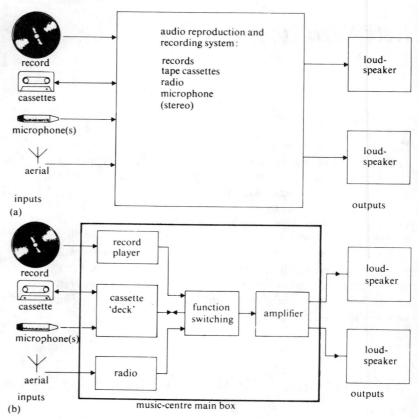

Figure 1 (a)   A simple system diagram for an audio system: (b) a system diagram for an audio system showing the subsystems

(*i*):   The lines connecting boxes indicate a flow of information between them, and the direction of flow is shown by an arrow head.

(*ii*): These lines are conceptual links to aid design, and do not necessarily represent electrical wires.

(*iii*): Most electronic circuits require a source of energy to make them function. This source is referred to as the power supply, and is often omitted from the system diagram to avoid having to draw all the connecting lines which might otherwise confuse the diagram.

(*iv*): The input–output relationship of a block is called the *transfer function*, and it defines the behaviour of the block over some specified range.

2

Circuit Diagrams

Circuit diagrams are used to define the electrical components used in a system. Unlike the conceptual blocks used in a system diagram, circuit diagrams use special symbols to represent each component. The circuit diagram for the torch is shown in Figure 2. Here the battery has been replaced by its symbol shown in Figure 3.

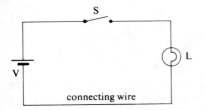

Figure 2   The circuit diagram for an electric torch

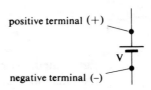

Figure 3   Circuit symbols for a voltage source

The complexity of todays digital integrated circuits often makes it very difficult to draw circuit diagrams for some of them. However, it is also unnecessary, because the operational characteristics of these devices are very well defined. What we need to know is where the connections are made. Hence a device like a microprocessor is drawn simply as a rectangular box with all the connection pins labelled.

The mixture of circuit symbols and boxes ensures that the details of the circuit are presented in a clear but concise way.

CHAPTER 2

Combinational Logic

Digital methods play a vital part in modern electronics, from computers to telecommunications, washing machines and stereo radio tuners. New applications in industrial, domestic and measuring equipment are constantly being found, so it is obviously important to understand the underlying ideas.

What does digital mean? You may think it has something to do with the 'digital displays' which show you a numerical answer on a pocket calculator or a digital voltmeter, but this is only a small part of the story. Much more important is the fact that the internal workings of digital devices depend on electrical effects which can be described by the so called 'binary digits' 0 and 1. These often correspond to two voltage levels, so that a typical digital signal, as shown in Figure 4, jumps back and forth between the two levels to represent a succession of 0s and 1s. I will illustrate how it is possible and useful to work with just 0s and 1s by the simple example of an industrial interlock, before going on to more general ideas. This will also show the link with 'logic', the systematic reasoning methods we use when trying to solve problems.

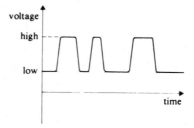

Figure 4   *A digital waveform*

A simple interlock

An 'interlock' is a common industrial requirement; it prevents a machine operating unless various safety conditions are satisfied.

For example, an automatic lathe may require the following three 'input' conditions, which can be labelled **A, B** and **C,** to be satisfied before the required 'output' **P** is obtained:

Input conditions

A, the workpiece is in position;
B, adequate lubricant is available;
C, a safety guard is in position.

Output requirement

P, the mains switch will turn on the machine.

If (binary) digital electronics is to be used to design such an interlock, we can recognize four separate steps:

Step 1: Express the possible input and output conditions in binary form.

Step 2: Tabulate the input and output conditions in a 'truth table' which specifies the interlock requirements.

Step 3: Convert the input and output conditions into *electrical* form by means of suitable transducers which produce voltages representing the binary digits.

Step 4: Find an electronic device with binary electrical inputs and outputs which are the same as the 'truth table' in Step 2.

Step 1: Binary inputs and outputs

If the input conditions (**A, B** and **C**) and the output requirement (**P**) are to be expressed in binary form, they must be phrased so that there are only two alternative possibilities in each case. These alternatives can then be labelled 0 and 1. There is usually more than one way of phrasing the conditions to give clear alternative possibilities; thus, condition **A** could be expressed in the form of the following questions:

'Is the workpiece in position?' Answer: YES or NO.

or

'Where is the workpiece; is it in or out of position?' Answer: IN or OUT;

or as the statement

'The workpiece is in position', which can be TRUE or FALSE.

The alternatives YES, IN or TRUE describe the same state of affairs, and would usually be represented by 1. The alternatives

NO, OUT or FALSE would then correspondingly be represented by 0. However, this is a matter of choice—the important things are that the condition should be carefully phrased so that there are only two possible choices, and that the decision as to which shall be represented by 1 and which by 0 should be clearly understood. We can then use the shorthand $A = 1$ to represent the situation with the workpiece in position, and $A = 0$ for the situation where it is absent. (This is to use the notation of *Boolean algebra*, named after the nineteenth century mathematician George Boole. A is described as a 'logical variable', which can only take two values, say 0 and 1.)

Boolean algebra

There are many other practical situations where two clear alternatives can be recognized: a lamp can be ON or OFF, a toggle switch can be UP or DOWN, a machine can STOP or GO, and so on. These are all described as two-*state* situations, and can easily be translated into the corresponding binary symbols 0 and 1. If they are expressed as statements which can only be TRUE or FALSE, this links the world of practical affairs with the formal language of 'logic' (which was studied long before electronics). Even continuously varying quantities like liquid level or voltage can be given a simple two-state description as HIGH or LOW in relation to some specified value – is the level ABOVE or BELOW a marked line or pointer reading? You will see that there is then a problem about values near the dividing line; something has to be specified about the precision of measurement and, particularly if the levels are fluctuating randomly, there will be uncertainty about some readings. This point must be borne in mind, and will recur. It is relevant to the second of the interlock conditions, **B,** which is not satisfactorily phrased as it stands; what does 'adequate lubricant available' mean? It can either refer to the *flow* of lubricant over the work, required to be more than a specified number of litres min^{-1}, or to the *level* of lubricant in a storage tank, required to be above some specified minimum. By setting these specified quantities sufficiently above the anticipated danger levels, a safety margin could be provided, allowing for random fluctuations in the flow or level. All these questions would have to be considered in the particular practical case in order to turn condition **B** into a simple binary statement $B = 0$ or $B = 1$.

How could condition **C** (safety guard in position) and output **P** (mains switch turns on machine) be phrased so as to provide clear binary alternatives.

Possible phrases are:

> The safety guard is in position TRUE (1) or FALSE (0)

or

> Is the safety guard in position YES (1) or NO (0).

Does the mains switch turn on the machine YES (1) or NO (0).

or

The mains switch turns on the machine TRUE (1) or FALSE (0).

Once the conditions have been phrased as binary alternatives, the designer can then move on to the second step.

Step 2: The truth table

The required relationship between the inputs **A, B** and **C** and the output **P** can now be tabulated in a 'truth table', which specifies the required interlock properties. Each possible combination of input conditions must be considered in turn. Thus, if the workpiece is *not* in position (**A** = 0), there is *not* adequate lubricant (**B** = 0) and the safety guard is *not* in position (**C** = 0), we require that the mains switch should *not* turn the machine on (**P** = 0). This can be written

C	B	A	P
0	0	0	0.

(The order of the terms **A, B** and **C** on the left does not matter, but you will see later why it is convenient to write them from right to left in this way.) If only one or two of the input conditions are satisfied, we still require **P** = 0; thus, if the workpiece is in position and there is adequate lubricant, but the safety guard is not in position, this can be written

C	B	A	P
0	1	1	0.

Only when *all three* input conditions are met, should the mains switch turn the machine on:

C	B	A	P
1	1	1	1.

A *complete* description of the required interlock properties involves a complete list of all possible combinations of inputs with the required output; the interlock will not be working properly if any of the input combinations gives the wrong output.

You will see that with three input conditions there are eight possible input combinations, only one of which gives **P** = 1. The required truth table is shown in the margin.

There is no reason why the rows should be listed in this particular order, and in fact as each row represents one acceptable combination of inputs and outputs there would be no change of

C	B	A	P
0	0	0	0
0	0	1	0
0	1	0	0
0	1	1	0
1	0	0	0
1	0	1	0
1	1	0	0
1	1	1	1.

7

meaning in the truth table if the rows were rearranged as long as all eight possibilities were included. It is, however, usual to write truth tables with the possible input combinations arranged in the systematic way shown here to avoid missing any, so it pays to learn the system. For the moment you can regard this as just a particular pattern, but after reading the next section of this chapter you will see how it is linked to number representation.

states

Each separate binary input condition has been described as being in one of two *states*. Two inputs can therefore be in one of four states, three inputs, as here, in one of eight. Each row of the truth table corresponds to one input state.

Now we would like to indicate some of the ways in which this interlock example could be extended. As considered so far there are three input conditions and one output condition, but of course there could be more of each. Suppose for example there were a fourth input condition:

D, a maintenance key is inserted,

which has the effect of overriding the other three conditions, so that when **D** = 1 (the key *IS* inserted, the condition is TRUE) the machine can be started (**P** = 1) whether the other conditions (**A, B, C**) are satisfied or not. This might be required by a maintenance man wanting to test the machine under abnormal conditions. The extra condition has the effect of doubling the number of rows in the truth table to 16; **D** = 0 the rest of the table is as before, when **D** = 1 there are a further eight rows all giving **P** = 1:

D	C	B	A	P	
0	0	0	0	0	
0	0	0	1	0	
0	0	1	0	0	
0	0	1	1	0	
0	1	0	0	0	as before
0	1	0	1	0	
0	1	1	0	0	
0	1	1	1	1	
1	0	0	0	1	
1	0	0	1	1	
1	0	1	0	1	
1	0	1	1	1	
1	1	0	0	1	additional rows
1	1	0	1	1	
1	1	1	0	1	
1	1	1	1	1.	

(You will see now why the input conditions are written from right to left, so that an extra condition could be added without disturbing the standard input pattern. However, when the number of inputs is known from the outset, it is quite usual to write **A, B, C, D** from left to right. It does not change the meaning of the truth table as long as the input columns are correctly identified with the corresponding conditions.)

In this particular case, where **P** = 1 for **D** = 1, whatever the values of **A, B** and **C** it is convenient *shorthand* to write

D	**C**	**B**	**A**	**P**
1	X	X	X	1.

to replace the last eight rows (where X can be read as 'don't care'), giving the truth table below:

D	**C**	**B**	**A**	**P**
0	0	0	0	0
0	0	0	1	0
0	0	1	0	0
0	0	1	1	0
0	1	0	0	0
0	1	0	1	0
0	1	1	0	0
0	1	1	1	1
1	X	X	X	1.

But it must be emphasized that this is only to save writing when it happens that the output does not depend on some of the inputs. A truth table is essentially an exhaustive statement of the output for *all* possible input combinations, so it must have (or imply) eight rows for three inputs, sixteen rows for four inputs, and so on.

From these examples, with 3, 4, 5 ... inputs giving 2^3, 2^4, 2^5 ... rows you will see that the general rule is:

> n input conditions lead to 2^n rows in the truth table. In other words, for n two-state input conditions there are 2^n input states.

There could equally well be more *outputs*; for example, it might be useful to have indicator lamps to show which of conditions **A, B** and **C** was not satisfied if the machine would not start. Thus, there could be an output

> **Q:** 'no workpiece' indicator lamp on,

with the requirement

> **Q** = 1 (lamp ON), when **A** = 0 (workpiece not in position)

and

> **Q** = 0 (lamp OFF), when **A** = 1 (workpiece in position).

Similarly, there could be outputs

> **R:** 'inadequate lubricant' lamp on,

and

> **S:** 'safety guard missing' lamp on,

with similar requirements

> **R** = 1, when **B** = 0; **R** = 0 when **B** = 1,
> **S** = 1, when **C** = 0; **S** = 0 when **C** = 1.

Finally there could be an 'UNSAFE' indication whenever these safety requirements were overriden by the maintenance key; that is, there would be an output condition

> **T:** 'UNSAFE' indicator on, with the requirement

> **T** = 0, when **D** = 0,
> **T** = 1, when **D** = 1, unless **A**, **B** and **C** are all = 1 at the same time.

The new truth table would have five outputs **P, Q, R, S** and **T** and the same four inputs, it would look like:

D	C	B	A	P	Q	R	S	T
0	0	0	0	0	1	1	1	0
0	0	0	1	0	0	1	1	0
0	0	1	0	0	1	0	1	0
0	0	1	1	0	0	0	1	0
0	1	0	0	0	1	1	0	0
0	1	0	1	0	0	1	0	0
0	1	1	0	0	1	0	0	0
0	1	1	1	1	0	0	0	0
1	0	0	0	1	1	1	1	1
1	0	0	1	1	0	1	1	1
1	0	1	0	1	1	0	1	1
1	0	1	1	1	0	0	1	1
1	1	0	0	1	1	1	0	1
1	1	0	1	1	0	1	0	1
1	1	1	0	1	1	0	0	1
1	1	1	1	1	0	0	0	0.

There are several things worth noting about this rather more complicated truth table. First, the four left-hand columns, the input conditions, contain a standard pattern of 16 rows which can be arranged in the same way for any four-input truth table, while the 16×5 binary digits in the right-hand output columns do not form a simple pattern; they contain the information that distinguishes this particular problem, and have to be filled in according to its particular requirements. Next, note that although the number of binary digits to be specified in these output columns *doubles* for each additional input condition, it is only *proportional* to the number of outputs; so there is a much faster increase in truth table size with the number of inputs than with the number of outputs. Finally, note that the introduction of the new output conditions has removed the possibility of a shorthand version of the last eight rows using Xs. Having specified the problem with a truth table, the designer now moves on to the third step.

Step 3: Electrical logic levels

The two steps taken so far have nothing to do with electronics; the conditions have simply been expressed in binary form and the required relation between inputs and outputs expressed as a truth table. The next step is to find a practical solution to the problem ('implement the truth table'). Even this does not *have* to involve electronics. You may be able to think of ingenious combinations of mechanical switches which would provide a practical interlock; it is often done that way. But this example is used to illustrate electronic logic (with no moving parts). The electrical represen- tation of 0 and 1 depends on recognizing two *states* in an electrical component – it could be that a current is flowing or not flowing in a wire, or a potential difference is present or absent between two terminals and so on.

There are various agreed standards, but let us consider a very common one, that used in 'TTL' devices, which will be discussed later in the chapter. The existence of a *'logic state'* 0 or 1 at any terminal depends on whether the voltage with respect to a common ground terminal is LOW or HIGH in comparison to a threshold voltage.

logic state

A typical high voltage might be 3.5 V, a typical low might be 0.2 V, but there can be quite a spread of allowed values. For example, any voltage in the range 2.4–5 V will be detected as HIGH and any in the range 0–0.4 V as a LOW. This is one advantage of digital methods, that exact binary decisions between 0 and 1, on which everything depends, do not require exact voltages.

It is convenient to associate HIGH voltage with TRUE, repre- sented by 1, and LOW voltage with FALSE, represented by 0 (this is the *'positive-logic'*, 'high-true' or 'active-high' convention), though the opposite is sometimes done. I will stay with the positive- logic convention at present. Now that we have described a method of representing logic levels electrically, we can draw a block diagram for the interlock system (Figure 5, p. 12).

positive logic

It will be necessary to find or devise *input* transducers which produce HIGH or LOW logic levels appropriate to the input conditions, and *output* transducers which convert logic levels into appropriate controls for the main switch and indicator lamps. The next task of the designer is to find a suitable electronic device, the fourth step.

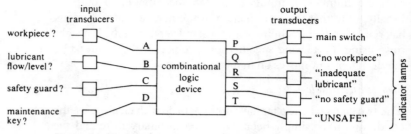

Figure 5   *Block diagram for the interlock system*

Step 4: Combinational logic devices

A typical device which might be used is an integrated circuit (IC) on a silicon chip in a plastic or ceramic package, as illustrated in Figure 6. It can be represented in a block diagram as in Figure 7. If it is a 'TTL' device the power supply required is 0 V and +5 V and the logic levels at the terminals must be HIGH or LOW as described above. For this particular application it must have (at least) four input and five output terminals, which can be labelled A, B, C, D and P, Q, R, S, T, corresponding to the logical variables **A, B, C, D, P, Q, R, S** and **T** discussed previously. The truth table arrived at in step 2 provides a specification for the required transfer function of this logic device.

Figure 6   *An integrated circuit*

Thus if the device is to implement the four input five output truth table, the application of input voltages corresponding to 1010 to terminals A, B, C and D, respectively, must produce output voltages corresponding to 00100 on P, Q, R, S and T. If the input is changed to 0100 the output must change to 01010 and so on. Many standard integrated circuits exist from which the required function can be chosen or built up. There are two distinct approaches.

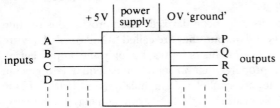

Figure 7   Block diagram of an integrated circuit

(a) ROMs

The most direct approach is to use a ROM, a *'read-only memory'*. There are various types, but each particular design is characterized by the number of inputs and outputs available; for example, a common type has ten inputs and eight outputs. But the transfer function linking outputs with inputs, the truth table, is not defined until the user chooses it in some way, either when it is ordered or after it is received.

read-only memory (ROM)

In both cases the effect is the same – the required truth table is stored on the chip, so that the outputs in a particular row of the truth table are obtained (for example, 00100) when the corresponding combination of inputs (for example, 0101) is provided. This is referred to as 'reading' the contents of the memory in that row, just as though you were reading from the written truth table. As the truth table stored on the chip is not changed in normal operation, this explains the name 'read-only' memory. The ROM method of providing combinational logic involves no further logic design work once the required truth table has been established and the ROM chosen. However, there are situations where a ROM may not be convenient, as will be discussed in later chapters.

(b) Gates

The alternative approach, used before cheap ROMs were available is to build up the required truth table from smaller components with certain standard logic properties built into them. For example, consider first the original three-input, one-output version of the interlock (page 7), for which the truth table was:

C	B	A	P
0	0	0	0
0	0	1	0
0	1	0	0
0	1	1	0
1	0	0	0
1	0	1	0
1	1	0	0
1	1	1	1.

13

logic function

gates

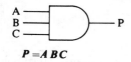

P =A B C

Figure 8   AND gate

inversion
inverter

This particular form of truth table, where **P** = 1 *only* when **A** AND **B** AND **C** are *all* equal to 1, describes the *'logic function'* called 'AND'.

Various circuits have been devised which provide this relation between inputs and output; they are called 'AND *gates'*. They are made with various numbers of inputs, but only one output, always with the same property that *all* inputs must equal 1 to give an output of 1. They are symbolized graphically as in Figure 8. In the notation of Boolean algebra, the AND function may be written:

P = **A.B.C,** or simply **ABC**. (This is *not* normal multiplication, but a special notation.)

Now consider the relationship between input **A** and output **Q** described above, which was that an indicator light was on (**Q** = 1) when the workpiece was absent (**A** = 0), and off (**Q** = 0) when the workpiece was present (**A** = 1). This relationship is provided by a one-input, one-output device with the property that the output is the opposite binary value to the input, a property called *inversion*. Such a device is called an *inverter*, and is symbolized as in Figure 9. In the notation of Boolean algebra, this property may be written

$$\mathbf{Q} = \mathbf{\bar{A}}.$$

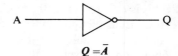

Figure 9   Inverter $Q = \bar{A}$

Several other standard logic functions and their corresponding symbols and notation will be introduced in a more detailed discussion of truth tables (pp 23–30). I hope it is already plausible that by suitably combining such standard functions, as provided by standard gates, truth tables of any complexity can be built up. In fact, there are well-established techniques for doing this, although the availability of ROMs has made them less necessary to learn. More complex integrated circuit components (including ROMs) are in fact built up in their internal construction from gates. Conventional abbreviations are:

SSI

SSI (small-scale integration) for integrated circuits fabricated with less than 12 gates,

MSI

MSI (medium-scale integration) 12–100 gates,

LSI

LSI (large-scale integration) more than 100 gates, and perhaps

VLSI

VLSI (very-large-scale integration) more than 10000 gates.

Terminology

This now completes the introductory review of combinational logic provided by the interlock problem. Before discussing the various steps in the solutions in a more general way, a few common terms with their explanations are listed below.

Bit: The normal abbreviation for a *binary digit*, 0 or 1. Whenever a group of 0s and 1s is written down, say in a truth table, or is present in electrical form, say as the voltage levels on the input terminals of a ROM, it can be described as a group (or string) of bits.

bit

(Binary) word: Just as the 26 letters of the alphabet can be grouped to form words with all sorts of meanings and uses, so the only two symbols we have available in digital electronics, the bit symbols 0 and 1, can be grouped to form (binary) *words* with various meanings and uses. In a particular digital system like a computer, it is usual to reserve the term *word* for a standard number of bits. Typical standards are words of 8, 16 or 32 bits.

binary word

Byte: This is a term like word defined for a particular system, but usually means a group of eight bits.

byte

Code: A group of bits used to specify one of several choices according to some agreed rules. You have seen how one particular set of interlock input conditions could be described by writing say

code

$$\begin{array}{cccc} \mathbf{D} & \mathbf{C} & \mathbf{B} & \mathbf{A} \\ 0 & 1 & 1 & 0. \end{array}$$

The group of bits 0110 may then be called a 4-bit input *code* word specifying one of the 16 possible combinations of inputs for this problem. As you will see in the next section, a 4-bit code can equally well describe one of (up to) 16 choices of any kind, provided we have agreed what the rules are for translating the group of bits into what it represents.

Address: A memory, such as a ROM, can be pictured as a number of locations into which bits have to be placed to build up the truth table of the particular problem. Indeed, such locations physically exist on a microscopic scale on the chip. The input code word which identifies a particular row of the truth table, the particular outputs corresponding to that input, is therefore often called an *address*.

address

Data: The bits stored in the ROM, which appear at the output in response to an address applied to the input, are called (output) *data*.

data

The input terminals of a ROM are therefore often labelled A_0, A_1, A_2 ... (for 'address') and the output terminals D_0, D_1, D_2 ...

(for 'data') as in Figure 10. A 10-input, 8-output ROM has a truth table which looks like this:

Row	A_9	A_8	A_7	A_6	A_5	A_4	A_3	A_2	A_1	A_0	D_0	D_1	D_2	D_3	D_4	D_5	D_6	D_7
0	0	0	0	0	0	0	0	0	0	0	0	1	1	0	0	1	1	0
1	0	0	0	0	0	0	0	0	0	1	1	0	1	1	0	1	1	1
2						etc.										etc.		
.	10-bit address codes standard pattern										8-bit (that is, one byte) data groups, depending on particular problem							
1022	1	1	1	1	1	1	1	1	1	0	1	0	1	1	0	1	1	0
1023	1	1	1	1	1	1	1	1	1	1	0	1	0	1	1	0	1	0

and can be described as an 8 Kbit (or 1 Kbyte) ROM, organized as $1\ K \times 8$ (or 'byte-organized'). Note the abbreviation K for 1024, not to be confused with the standard metric k for 1000.

bus

Bus: The group of wires connected to the input terminals and carrying addresses is called an 'address *bus*'; the group connected to the output terminals is similarly called a 'data bus'.

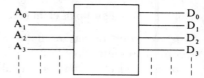

Figure 10

Binary inputs and outputs – coding

The interlock example has shown you one way in which inputs and outputs can be expressed in the binary form required for a combinational logic device. There are several other ways binary inputs and outputs can arise, and they form the subject of this section. There are inputs and outputs which may be classified as 'intrinsically binary' (as in the interlock), and there are others which are not, requiring to be coded into binary form in some way. In this second group are the binary inputs and outputs which represent numbers (as needed for calculators), other symbols like letters, or any other set of distinct items.

Intrinsically binary inputs and outputs

In the interlock case considered on p. 5, the description of inputs and outputs in binary form was straightforward, because they were

16

expressed as questions with clear YES/NO answers or statements which were clearly TRUE or FALSE; that is, they each had only two distinct states. A particular combination of input conditions was then represented by a group of bits forming an input *code* word which could be used as the address of a ROM. One bit describes the two possible states of one input, two bits describe the four possible states of two inputs, three bits the eight possible states of three inputs and so on, each additional bit doubling the size. An address code word picks out a particular state from all those possible for the given number of inputs. If there were n inputs, and therefore an n-bit code, there would be 2^n possible states. This is a general rule for binary coding – the maximum number of distinct states capable of being represented by an n-bit code is 2^n. It does not follow that they all have to be used. In the case of an *address* code for a truth table, say, all combinations of 0 and 1 are used because the logical transfer function must be specified for all possible inputs; in the particular interlock example all input combinations are possible. But in some cases not all input codes have any meaning in the particular conditions. of the problem, examples being provided by BCD codes in the next section.

Number representation

It is often necessary to find a binary code for input and output information which is *not* intrinsically in binary form. One of the commonest examples is *number representation*, the conversion of our familiar denary numbers (like 357, 35700, 3.57, 0.00357 – often loosely called decimal numbers) to and from binary form so that they can be handled by a computer but understood by humans. When you press a number key on a calculator, the electrical input produced must represent the number in some binary code as the internal circuits operate on binary voltage levels. The coding can be done in various ways, some of which make use of all 2^n values of an n-bit code, others of which do not.

Natural binary

The normal denary number system uses ten symbols 0, 1, 2 . . . 8, 9 arranged side by side with a 'positional weighting' system which changes their significance (or weight) according to their position. Thus,

$$397 = 3 \text{ hundreds} + 9 \text{ tens} + 7 \text{ units}$$
$$= 3 \times 10^2 + 9 \times 10^1 + 7 \times 10^0.$$

The positional weights are powers of ten. Note that the positions increase in significance from right to left (units, tens, hundreds . . .), the least significant being the zeroth power, and the nth from

17

natural binary

the right being the $(n-1)$th power of 10. The binary system uses only two symbols, 0 and 1. So what is called the *natural binary* number system, built up on the same basis as denary, has corresponding place weights which are powers of *two*:

$$101 \qquad\qquad = 1 \text{ four} + 0 \text{ twos} + 1 \text{ unit}$$
$$(\text{natural binary}) \quad = 1 \times 2^2 + 0 \times 2^1 + 1 \times 2^0$$
$$= 5 \text{ (denary)}.$$

This can be extended indefinitely, just like the denary system, by adding digits to the left representing 'eights' (2^3), 'sixteens' (2^4) and so on. The bit at the extreme right (representing units 2^0) is known as the *least significant bit'* (LSB), that at the extreme left (representing 2^{n-1} for n bits) the *'most significant bit'* (MSB).

LSB
MSB

A long natural binary number can be converted into denary simply by adding the appropriate powers of two:

$$1011011_2 = 1 \times 2^6 + 0 \times 2^5 + 1 \times 2^4 + 1 \times 2^3 + 0 \times 2^2 + 1 \times 2^1 + 1 \times 2^0$$
$$= 64 \qquad + \qquad 16 \ + \ 8 \qquad\qquad +2 \qquad +1$$
$$= 91_{10}$$

(Where it is necessary to avoid confusion, the subscripts 2 and 10 are introduced to indicate natural binary or denary, respectively.)

Similarly, a denary number can be converted into natural binary by looking for appropriate powers of two:

$$397_{10} = 256 + 128 \qquad\qquad\qquad +8 \quad +4 \qquad\qquad +1$$
$$1 \times 2^8 + 1 \times 2^7 (+ 0 \times 2^6 + 0 \times 2^5 + 0 \times 2^4) + 1 \times 2^3 + 1 \times 2^2 (+ 0 \times 2^1) + 1 \times 2^0$$
$$= 1 \qquad 1 \qquad 0 \qquad 0 \qquad 0 \qquad 1 \qquad 1 \qquad 0 \qquad 1_2$$

Another way of arriving at this is to keep dividing by two, and list the remainders, r, as shown alongside.

Reading the remainders r in the order shown by the arrow:

$$397_{10} = 110001101_2 .$$

2	397	
2	198	$r = 1$ LSB
2	99	$r = 0$
2	49	$r = 1$
2	24	$r = 1$
2	12	$r = 0$
2	6	$r = 0$
2	3	$r = 0$
2	1	$r = 1$
	0	$r = 1$ MSB

(This works because at the last division the original 397 has by now been divided by 2 nine times, that is by 2^9, and the answer is 0 r 1. So there is no 2^9 term, but 1×2^8 left over from the previous division, and so on.)

You have in fact already met the natural binary number system in the form of the truth table address code. If you look back at the truth tables you will see that the standard pattern of addresses simply corresponds to numbering the rows in the natural binary number system (including 0s at the left to make up a fixed number of bits). Thus, a 3-bit natural binary address code extends from 000 to 111, representing 0 to 7_{10}. Similarly a 4-bit natural binary address code extends from 0000 to 1111, representing 0 to 15_{10}. This is also referred to as an 8:4:2:1 code, to indicate the weights of the four bits, the successive powers of two the positions represent.

BCD code

Binary-coded decimal (BCD) codes

As you have seen, denary numbers of any size can be coded into natural binary form provided enough bits are used. However, there

18

is no simple link between the separate symbols of the denary number and the resulting binary code; you have to work out the arithmetic as in the previous section.

An alternative approach is to replace *each* denary digit in turn by a 4-bit binary code. There are ten denary digits; four is the smallest number of bits n that give $2^n > 10$. This means that there will be six of the possible 16 values of the 4-bit code that have no meaning; so this is an example of the cases mentioned above where not all values of a code mean anything. The most natural BCD code is an '8:4:2:1' one, which uses the first ten 4-bit natural binary codes to represent 0–9.

This may be clearer if the corresponding 'natural binary' and 'BCD' codes for denary 0–15 are written side-by-side:

denary	natural binary	8:4:2:1 BCD
0	0000	0000
1	0001	0001
2	0010	0010
3	0011	0011
4	0100	0100
5	0101	0101
6	0110	0110
7	0111	0111
8	1000	1000
9	1001	1001
10	1010	0001 0000
11	1011	0001 0001
12	1100	0001 0010
13	1101	0001 0011
14	1110	0001 0100
15	1111	0001 0101.

Note that for denary numbers 0–9, with only one digit symbol, the corresponding BCD code has only 4 bits, but for 10–15, with two digits, the BCD code has 8 bits. Both systems can be extended indefinitely. A 5-bit natural binary code can represent 0–31, 6-bit 0–63 and so on, while BCD codes simply require 4 bits per denary digit. As you have seen, $397_{10} = 110001101_2$ in natural binary. For BCD coding simply take each digit separately:

$$\begin{array}{ccc} 3 & 9 & 7 \\ 0011 & 1001 & 0111. \end{array}$$

What are the natural binary and 8:4:2:1 BCD codes for denary 420? The natural binary number is found from the division by 2 procedure, shown here.

Hence 420 = 10100100 natural binary. The BCD code is found by taking each decimal digit separately, therefore

420 = 0100 0010 0000 BCD.

2	420		
2	210	r = 0	LSB
2	165	r = 0	
2	52	r = 1	
2	26	r = 0	
2	13	r = 0	
2	6	r = 0	
2	3	r = 1	
2	1	r = 0	
	0	r = 1	MSB

You will have noticed that the natural binary code for large denary numbers is shorter than the corresponding BCD code, but that it is much easier (for denary-educated humans) to translate quickly to and from BCD. Computers in their internal operations use codes like natural binary, but at the points of contact with people (keyboards, displays) often use BCD codes. For various computational reasons the codes used are not necessarily the ones described here; the 16 values of a 4-bit code used to represent denary 0–15, or the ten of them selected to represent 0–9, need not be in the natural binary (8:4:2:1) order I have given here. Other choices are possible and are used, but I will not go into them any further here. The important principle is to distinguish between the use of n bits to represent 2^n denary numbers directly, and the less economical, but sometimes more convenient, BCD method of using 4 bits at a time to represent separate denary digits.

Octal and hexadecimal

There is one more useful idea to be introduced about binary coding of numbers. This is concerned with the reverse problem of quickly reducing a long string of binary digits to a more manageable shorthand form for easy reference or checking by people. Of course the binary string in question can always be treated as natural binary, and translated into denary, but it is not possible to do this very quickly. Two alternatives are commonly used; these are called *octal* and *hexadecimal*.

octal
hexadecimal

Suppose the initial binary string is 11010110. To translate this into 'octal', the binary string is divided into groups of three bits (counting from the LSB at the right), that is 11, 010, 110. Each group is replaced by the corresponding digit 0–7 (according to the natural binary code, which only extends as far as 7 for three bits), that is

$$(0)11, 010, 110$$
$$3 \quad 2 \quad 6_8.$$

The subscript 8 indicates that this is an octal number not a denary one. (The denary equivalent of 11010110 is $128 + 64 + 16 + 4 + 2 = 214_{10}$.) The idea of hexadecimal (commonly called 'hex') is to extend this principle to *four* bits at a time, that is 1101, 0110. In order to replace each group of four bits by a single symbol (from 0000 to 0101) 0–9 can be used as usual, but what about the remaining 4-bit groups (equivalent to 10–15 in denary)? New single symbols are needed: it is normal to use the letters A–F.

Thus

	denary	Hexadecimal	octal
0000	0	0	0
0001	1	1	1
0010	2	2	2
0011	3	3	3
0100	4	4	4
0101	5	5	5
0110	6	6	6
0111	7	7	7
1000	8	8	10
1001	9	9	11
1010	10	A	12
1011	11	B	13
1100	12	C	14
1101	13	D	15
1110	14	E	16
1111	15	F	17
10000	16	10	20
10001	17	11	21.

So $\underline{1101}\ \underline{0110}$

can be written $\quad D \quad 6_{16}$

in hex, again including the subscript 16 to emphasize that this is not denary. The only object of these conversions is to arrive at a convenient shorthand; it is much easier to recognize or recall short strings of octal or hex digits than long strings of 0s and 1s.

Other symbols and distinct items

Another common requirement is a code for letters and other characters as well as numbers, so that messages can be sent in binary digital form. An important example is the *ASCII* code (American Standard Code for Information Interchange), used for feeding information into computers from keyboards, operating alphanumeric displays, printers, and so on.

ASCII

This code uses 7 bits to represent $2^7 = 128$ characters, consisting of upper and lower case letters, numbers, punctuation marks and special symbols such as £ and $. The codes 0–31 are used for special non-printing control characters such as carriage return. The characters represented by the codes 32–128 are listed in Table 1 (p. 22). The seven bits are labelled b_1–b_7, and b_1 is the LSB.

There are two more columns (with $b_7 b_6 b_5 = 000$ and 001) which have not been included as they represent messages requiring detailed explanation. Note that numbers are included in the column $b_7 b_6 b_5 = 011$; the other four bits (b_1–b_4) form the 8:4:2:1 BCD code you have already seen. In practice, an eighth *'parity' bit* b_8 may be added at the MSB end, chosen to make the number of 1s odd (say) for all symbols. (It could alternatively be chosen to make the number of 1s even, but the choice must be the same for all symbols.)

parity bit

Table 1 In this table the seven bits are labelled b_1–b_7, starting from the LSB.

b_4	b_3	b_2	b_1	Row	2	3	4	5	6	7
					b_7 = 0	0	1	1	1	1
					b_6 = 1	1	0	0	1	1
			Column		b_5 = 0	1	0	1	0	1
0	0	0	0	0	SP	0	@	P	`	p
0	0	0	1	1	!	1	A	Q	a	q
0	0	1	0	2	"	2	B	R	b	r
0	0	1	1	3	#	3	C	S	c	s
0	1	0	0	4	$	4	D	T	d	t
0	1	0	1	5	%	5	E	U	e	u
0	1	1	0	6	&	6	F	V	f	v
0	1	1	1	7	'	7	G	W	g	w
1	0	0	0	8	(	8	H	X	h	x
1	0	0	1	9	)	9	I	Y	i	y
1	0	1	0	10	*	:	J	Z	j	z
1	0	1	1	11	†	;	K	[	k	{
1	1	0	0	12	,	<	L	\	l	¦
1	1	0	1	13	–	=	M	]	m	}
1	1	1	0	14	.	>	N	^	n	~
1	1	1	1	15	/	?	O	—	o	DEL

Thus, the code for 3 (0110011) has four 1s, and so requires an additional 1 to give it odd parity; the code for R (1010010) with three 1s has odd parity already, so the additional bit would be 0. The reason for doing this is that if a single error occurs when the code is first produced, or during transmission from one place to another, so that a 1 is substituted for a 0, or 0 for a 1, the parity would change to even. So the presence of a single error would be detected if the number of 1s in each character were checked. This is a simple example of an *error-detecting* code: with more additional bits, an error can be located, more errors detected or limited numbers of errors *corrected*.

Obviously other codes than ASCII are possible, but there is no difference in principle: all that is needed is a table listing the symbols and the corresponding code. A binary code can equally well be associated with names of individuals, objects in a catalogue, and so on. A simple example is the display of coloured lights you may have seen in a large store to call staff members without disturbing customers. The 'code book' for such a system might read

	red	yellow	green	
Mr. Smith	0	0	1	(0 = light OFF)
Dr. Jones	0	1	0	(1 = light ON)
Ms. Robinson	0	1	1.	

It is important to note that the code word 000 (all lights off) must be assigned like any other. The most natural choice would no doubt be to assign it to 'nobody'; that is, if no lights are on, nobody is being called. But this is not the only possible choice; another option might be

| nobody | 1 | 0 | 0 |
| duty electrician | 0 | 0 | 0, |

which would have a 'fail-safe' feature. One red light would simply indicate that the system was working; if it went out it might be a fault requiring attention. The point of these simple examples is only to illustrate that coding starts as an *arbitrary* assignment of a string of bits to the various items to be coded, but it may be possible to build in safeguards by choosing the code carefully. The only general limitation is the 2^n rule – with n bits only 2^n distinct code words can be formed, so this is an upper limit on the number of distinct items which can be coded.

Binary input codes can also be used to represent analogue inputs, and binary outputs converted to analogue outputs. These will be described in Chapter 4. The important thing to remember is that we decide what the binary code represents, whether it be characters, people, numbers or analogue signals.

Truth tables; Boolean notation

The truth table has already been introduced as the specification of the logic required in a particular problem, and as the transfer function of the electronic black box needed to implement it. It has been emphasized that if a ROM is used the logic design problem is essentially solved once the truth table is written down (provided only that a large enough ROM is available). Let us now consider the development of truth tables of increasing complexity, introducing new terminology and notation for some standard logic functions. These are not essential for the ROM implementation (there is no need to name the logic function as long as the truth table is available), but are needed for the development of the alternative gate implementation.

Single output truth tables

One input

Starting with the very simplest case, consider a one-input, one-output combinational logic device, and its corresponding truth

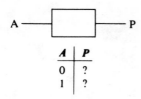

A	P
0	?
1	?

Figure 11

table (Figure 11). How many different possible truth tables are there? With one input there are only two possible input states, so there are two addresses or rows in the truth table. The output column can therefore be completed in four different ways:

A	P_0	or	A	P_1	or	A	P_2	or	A	P_3
0	0		0	0		0	1		0	1
1	0		1	1		1	0		1	1

The first and fourth of these are of little practical significance. In Boolean notation, they indicate $P_0 = 0$ or $P_3 = 1$; that is, the output logic level does not depend on the input at all, but is stuck at 0 or 1, respectively. The second of the four possibilities is almost as trivial: it implies that the output is the same as the input, which can be written $P_1 = A$. As far as logic levels are concerned, the terminals **A** and **P** might as well be connected together by a single wire. However, when the more detailed interconnection properties of logic elements are considered a $P = A$ device is not necessarily useless. Its electrical characteristics might provide a *buffer* between a source of logic signals having limited power and a further logic device needing more power to drive it than the source can provide. The graphical symbol for the buffer is shown in Figure 12.

The third possibility is the only one normally recognized as a useful logical function, the *inverter*. Whatever the value of **A** (0 or 1), **P** is always the other binary value (1 or 0) variously referred to as NOT-A, the COMPLEMENT of A, or the INVERSE of A. The Boolean notation for this is

$$P = \overline{A}.$$

Note that inverting twice restores the original value; that is

$$\overline{P} = \overline{\overline{A}} = A.$$

The graphical symbol is shown in Figure 13, where it is the small circle that implies inversion; it is used generally to mean this, as you will see. The triangle is the standard amplifier symbol, representing the fact that an inverter may also act as a buffer.

buffer

inverter

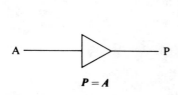

Figure 12

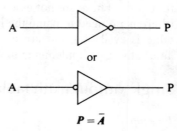

Figure 13

Two inputs

Now consider a two-input one-output device and its corresponding truth table (Figure 14). With two inputs there are $2^2 = 4$ input states, addresses or rows in the truth table. The output column can therefore be completed in $2^4 = 16$ different ways. I have listed the separate possibilities and numbered them from 0 to 15 (as though the **P** column were a 4-bit natural binary number):

A	B	possible forms for **P**
0	0	0 0 0 0 0 0 0 0 1 1 1 1 1 1 1 1
0	1	0 0 0 0 1 1 1 1 0 0 0 0 1 1 1 1
1	0	0 0 1 1 0 0 1 1 0 0 1 1 0 0 1 1
1	1	0 1 0 1 0 1 0 1 0 1 0 1 0 1 0 1
Column number:		0 1 2 3 4 5 6 7 8 9 10 11 12 13 14 15

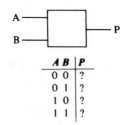

Figure 14

(This is *not* a 16-output truth table, but presents 16 possible choices for a single output. Any one combinational problem, or one combinational device, with two inputs and one output will have only one **P** column in its truth table, chosen from the sixteen listed here.) Once again quite a lot of these can be dismissed as of little practical significance.

For example, columns 0 and 15 show that the output **P** is independant of the inputs. **A** and **B**, whilst columns 3 and 12 depend on **A** only and 5 and 10 on **B** only. This now leaves ten cases where the output is dependant upon some combination of the two inputs.

One you have met before (in a three-input version) in the interlock problem is the AND function, taking practical form in an AND-gate. You will recognize it in column 1; the output is 1 only when both input **A** AND input **B** are 1. (For more inputs this becomes: output is 1 only when *all* of inputs **A** AND **B** AND **C** ... are 1.) It is written

$$\mathbf{P} = \mathbf{A.B}, \text{ or simply } \mathbf{AB},$$

and symbolized as in Figure 15. Another important case is the *OR*-function, taking practical form in an OR-gate, and found in column 7. It is the solution to a logic problem like: an entrance door can be opened (**P** = 1) if either of two watchmen, or both, have inserted their personal identifying keys in corresponding locks (**A** = 1 or **B** = 1 or both). The output is 1 when *any* or all of **A** OR **B** (OR **C** ..., for more inputs) are 1. It is written

$$\mathbf{P} = \mathbf{A} + \mathbf{B},$$

and symbolized in Figure 16.

Three points that often cause confusion are worth mentioning here:

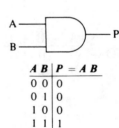

A B	P = A B
0 0	0
0 1	0
1 0	0
1 1	1

Figure 15

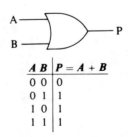

A B	P = A + B
0 0	0
0 1	1
1 0	1
1 1	1

Figure 16

25

(*i*): Be careful not to read **A** + **B** as **A** *and* **B**, or **A** *plus* **B**; in this Boolean notation it means **A** OR **B**, and is not the same as addition. Similarly, **A.B**, the notation for **A** AND **B**, is not the same as multiplication.

(*ii*): Some English sentences containing 'and' in fact imply the OR-function: to say that 'students and teachers will find this book useful' means a reader will find the book useful (**P**) if the reader is a student (**A**) OR a teacher (**B**), in the logic sense of OR.

(*iii*): There are mathematical objections to the use of '.' (the product symbol) for AND and '+' (the addition symbol) for OR, although this was Boole's original notation, and it is normally used in electronics.

A preferred notation which you may sometimes see is

$$A \cap B,$$

for AND, and

$$A \cup B,$$

for OR, sometimes called 'cap' ($\cap$) and 'cup' ($\cup$) notation.

Returning to the truth tables, one can see how six more of the possible output columns can be described by a combination of AND or OR with INVERSION. Two of them have names: column 14 is the inverse of column 1 (AND) and is called NAND; column 8 is the inverse of column 7 (OR) and is called NOR. There is something more to be said about this; consider the truth table shown in Figure 17, and the various ways NAND can be described, in Boolean notation, graphical symbols or in words. The first three output columns simply repeat cases you have already seen, to make it easier to follow the argument. The third column, AND, might represent, for example, a simple interlock, allowing a car to start only if (**A**) the doors are locked AND (**B**) seat belts of occupied seats are fastened. Then the two final columns (which are both the same) could represent the condition for a warning light to come on when the conditions are not met. You can look at this in two ways, either as:

	A	B	P
NAND	0	0	1
	0	1	1
	1	0	1
	1	1	0

	A	B	P
NOR	0	0	1
	0	1	0
	1	0	0
	1	1	0

$\overline{AB}$, the light comes on if it is NOT the case that both conditions **A** AND **B** are satisfied,

or as

$\overline{A} + \overline{B}$, the light comes on if either condition **A** is NOT satisfied OR condition **B** is NOT satisfied.

The truth table and the verbal descriptions both indicate that these amount to the same thing: the NAND function. NAND can be seen either as AND followed by NOT, or as NOT for each

26

input followed by OR; the two alternative symbols shown in Figure 17, with circles for inversion 'before' or 'after' (that is, to left or right of) the simple gate symbols, convey this equivalence. Thus, NAND and NOR are obtained either by inverting the output of AND and OR, or by inverting *both* the inputs of OR and AND.

A	B	$\overline{A}$	$\overline{B}$	AB	$\overline{A}\overline{B}$	$\overline{A} + \overline{B}$
0	0	1	1	0	1	1
0	1	1	0	0	1	1
1	0	0	1	0	1	1
1	1	0	0	1	0	0
		NOT-A	NOT-B	AND	NAND	

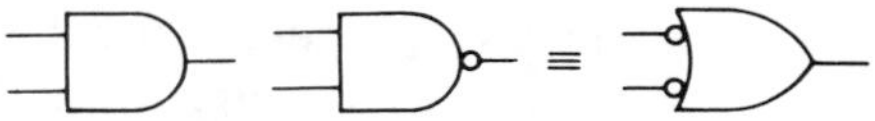

Figure 17

Inverting just *one* of the inputs provides a description for columns 2, 4, 11 and 13. If you examine the previous truth tables you will see that $A\overline{B}$ describes column 2 and $A + \overline{B}$ describes column 11:

A	B	$\overline{B}$	$A\overline{B}$	$A + \overline{B}$
0	0	1	0	1
0	1	0	0	0
1	0	1	1	1
1	1	0	0	1

(col. 2) (col. 11).

We are now left with just two columns unaccounted for (6 and 9), the only ones with two 0s and two 1s which are not simply the inputs or their inverses. Column 6 is given a special name, *EXCLUSIVE-OR (XOR)* to convey that the output is 1 if **A** or **B** is 1, but *not both*. It is written:

$$P = A \oplus B,$$

and symbolized as in Figure 18.

Column 9 is simply the inverse of this, XNOR, which can be written:

$$P = \overline{A \oplus B} \text{ or } A \overline{\oplus} B:$$

A	B	P
0	0	1
0	1	0
1	0	0
1	1	1.

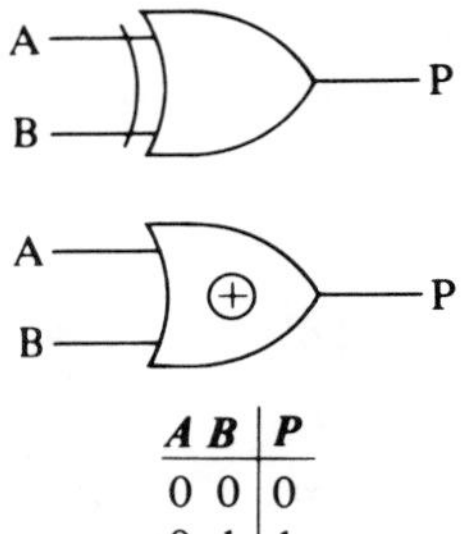

XOR

A B	P
0 0	0
0 1	1
1 0	1
1 1	0

Figure 18

27

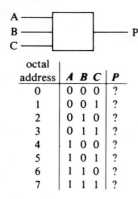

octal address	A	B	C	P
0	0	0	0	?
1	0	0	1	?
2	0	1	0	?
3	0	1	1	?
4	1	0	0	?
5	1	0	1	?
6	1	1	0	?
7	1	1	1	?

Figure 19

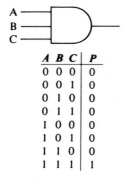

A	B	C	P
0	0	0	0
0	0	1	0
0	1	0	0
0	1	1	0
1	0	0	0
1	0	1	0
1	1	0	0
1	1	1	1

Figure 20

That completes the examination of all the 16 possible logic functions of the two variables **A** and **B**. As you have seen, not all of them are equally important; you should try to remember INVERT, AND, OR, NAND, NOR and XOR, as they will constantly recur.

More than two inputs

Extension to more inputs is straightforward, but increasingly laborious. For three inputs (Figure 19) there are $2^3 = 8$ input states or addresses, which can be conveniently numbered as octal addresses, and $2^8 = 256$ ways of filling in the output column. As before, there are some cases which do not depend on all three inputs ($P = 0, 1$; $A, B, C; \overline{A}, \overline{B}, \overline{C}; AB, \overline{A}B, BC \ldots; A + B, A + \overline{C} \ldots; A \oplus B, \overline{A} \oplus C \ldots$). The many more important cases where the output depends on all three inputs are of two kinds. First there are those which simply extend the definitions of AND, OR and their inverses which you have already met. The AND function for example has only a single 1 at the bottom of the output column; this was the case for the three-input interlock introduced on p. 00 (Figure 20). Any other truth table with a single 1 can be expressed as AND with one or more of the inputs inverted. For example, a telephone conversation can only take place (**P**) with the right relationship between the conditions

    A the caller's handset is on the rest;
    B the correct number is dialled;
    C the receiver's handset is on the rest.

The requirement is $A = 0$ AND $B = 1$ AND $C = 0$. The truth table is

A	B	C	P
0	0	0	0
0	0	1	0
0	1	0	1
0	1	1	0
1	0	0	0
1	0	1	0
1	1	0	0
1	1	1	0,

and this can be written

$$P = \overline{A}B\overline{C}.$$

Similarly, any truth table with a single 0 in the output column can be written as OR, possibly with one or more of the inputs inverted.

Then there is a large number of output possibilities with 2, 3 or 4 0s or 1s in the output column, not covered by any of the above.

For example, a car seat belt interlock, in more detail, might require that the car should only start (**P**) if the driver's seat belt is fastened (**A**), and (i) the front passenger seat is occupied (**B**) and the passenger seat belt is fastened (**C**), or (ii) the front passenger seat is unoccupied.

The truth table is now

A	B	C	P	
0	0	0	0	
0	0	1	0	
0	1	0	0	
0	1	1	0	⎤ Option (ii): if **B** = 0 (passenger
1	0	0	1	⎦ seat unoccupied), it does not matter
1	0	1	1	whether **C** is 0 or 1.
1	1	0	0	
1	1	1	1 ←	Option (i).

Option (i) alone could be written $\mathbf{P} = \mathbf{ABC}$; option (ii) could be written $\mathbf{P} = \mathbf{A\overline{B}}$ (from what I have explained previously). Because the requirement is $\mathbf{P} = 1$ if either (i) OR (ii) is satisfied they may be combined using the notation for OR: $\mathbf{P} = \mathbf{ABC} + \mathbf{A\overline{B}}$.

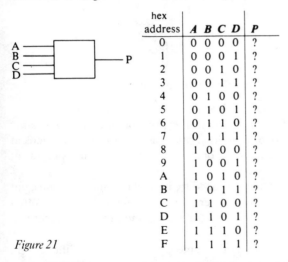

hex address	A	B	C	D	P
0	0	0	0	0	?
1	0	0	0	1	?
2	0	0	1	0	?
3	0	0	1	1	?
4	0	1	0	0	?
5	0	1	0	1	?
6	0	1	1	0	?
7	0	1	1	1	?
8	1	0	0	0	?
9	1	0	0	1	?
A	1	0	1	0	?
B	1	0	1	1	?
C	1	1	0	0	?
D	1	1	0	1	?
E	1	1	1	0	?
F	1	1	1	1	?

Figure 21

For four-inputs, one-output (Figure 21), the output possibilities become very numerous; there are $2^4 = 16$ input addresses (which can be conveniently labelled as hex addresses), and therefore $2^{16} = 65536$ ('64 K') ways of filling in the output column. There are now many cases which do not depend on all four inputs ($\mathbf{P} = 0, 1$; $\mathbf{A}, \mathbf{\overline{C}} \ldots$; $\mathbf{AB}, \mathbf{\overline{C}D}, \mathbf{\overline{A}} + \mathbf{C}, \mathbf{B} + \mathbf{D} \ldots$; $\mathbf{ABC}.\mathbf{B\overline{C}D}, \mathbf{A} + \mathbf{B} + \mathbf{D}$, $\mathbf{\overline{B}} + \mathbf{C} + \mathbf{D}, \ldots$ and so on), and even more which do ($\mathbf{ABCD}$, $\mathbf{A} + \mathbf{B} + \mathbf{C} + \mathbf{\overline{D}}, \mathbf{\overline{A}BC\overline{D}} + \mathbf{AB\overline{C}D}$, and so on.) You will have seen by now that each 1 in the output column corresponds to a term like

$A\bar{B}CD$ in the expression for **P**. The 16 terms like $A\bar{B}CD$ with or without inversion bars ($ABCD$, $\bar{A}\bar{B}\bar{C}\bar{D}$, $AB\bar{C}D$, and so on) are called *'minterms'*. So *any* truth table can be expressed in *'sum of*

sum of minterms

minterms' form by writing the minterms corresponding to each of the 1s in the output column and 'summing' them (in the logic sense, that is combining them by the OR function). Thus, for a truth table

A	B	C	D	P	
0	0	0	0	0	
0	0	0	1	1	←$\bar{A}\bar{B}\bar{C}D$
0	0	1	0	0	
0	0	1	1	0	
0	1	0	0	1	←$\bar{A}B\bar{C}\bar{D}$
0	1	0	1	0	
0	1	1	0	0	
0	1	1	1	0	
1	0	0	0	1	←$A\bar{B}\bar{C}\bar{D}$
1	0	0	1	0	
1	0	1	0	1	←$A\bar{B}C\bar{D}$
1	0	1	1	0	
1	1	0	0	0	
1	1	0	1	0	
1	1	1	0	0	
1	1	1	1	0,	

the Boolean expression for **P** may be written

$$P = \bar{A}\bar{B}\bar{C}D + \bar{A}B\bar{C}\bar{D} + A\bar{B}\bar{C}\bar{D} + A\bar{B}C\bar{D}.$$

This way of writing a Boolean expression does not necessarily give the most eonomical description, as you will see if you look at previous examples but it is one certain and comprehensive way of doing it.

There is a systematic procedure for making simplifications, but it need not be considered here. In electronic problems it is often sufficient to write down the truth table and proceed straight to a ROM.

For still more inputs the same principles apply, although the truth table becomes very tedious to write out. Combinations of the AND function $ABCDEF$ ... (with a single 1 at the bottom of the truth table), inversion (which can describe minterms with 1s anywhere else) and the OR function $A + B + C + D + E + F + $ ... (with a single 0 at the top of the truth table) provide a Boolean equivalent of any truth table.

Multiple output truth tables

We have already seen how an interlock truth table can be extended to describe a number of additional indicator lamp outputs. It is always possible that a problem may involve several outputs in this way. If the output requirements can be described separately in terms of the input conditions there is not much more to say; each output column is written down in turn as we have done for the single output. Frequently, however, the output columns are related to each other in some way, as in the following examples.

Binary arithmetic

When binary digits are used for number representation, it is important to be able to add them together. This is the basic operation in a digital computer, and can be readily extended to cover subtraction, multiplication and division.

Remembering the significance of binary numbers, the rules for addition are simple:

$$
\begin{array}{cccc}
0 & 0 & 1 & 1 \\
\underline{0} & \underline{1} & \underline{0} & \underline{1} \\
0 & 1 & 1 & 10.
\end{array}
$$

As in normal arithmetic, we can say for the answer to the last sum '0 carry 1'. That is, we write a 0 in the same column as the bits being added, and we 'carry' a 1 into the next column on the left. A device which performs this simple addition is called a *half adder*, with two inputs A and B, and two outputs S (sum) and C (carry). The required truth table is therefore

Note that $S = A \oplus B$ (the XOR function) and $C = A.B$ (the AND function).

In all columns of an extended addition sum except that at the right, there is the possibility of a digit carried from the previous columns as well as the digits to be added. Thus, for example, to add 101 and 111, we must write

half adder

A	B	C	S
0	0	0	0
0	1	0	1
1	0	0	1
1	1	1	0.

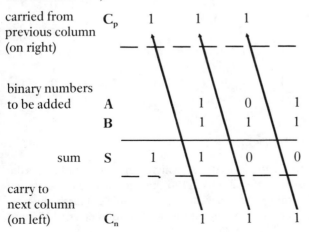

carried from previous column (on right) C_p 1 1 1

binary numbers to be added A 1   0   1
                              B 1   1   1

sum S 1   1   0   0

carry to next column (on left) C_n 1   1   1

Thus, in the right-hand column, we have $1 + 1$ giving $S = 0$ and the carry to the next column $C_n = 1$ (as before). This is then transferred to the next column to become C_p (carry from previous column) $= 1$, to be added to 0 and 1 to give $S = 0$, $C_n = 1$. Transferring again giving $C_p = 1$, to be added to 1 and 1 to give $S = 1$, $C_n = 1$. The final transfer gives $C_p = 1$ to be added to nothing so the left-hand term in $S = 1$.

full adder

A device which performs one column of this addition is called a *full adder* with three inputs A, B and C_p, and two outputs S and C_n. The full adder is a simple basic component for a computer. In fact more complicated devices called Arithmetic Logic Units (ALUs) are available which can perform a range of arithmetical operations.

Code conversion

The various outputs for a particular input address can often be regarded as a related group. A standard form of ROM for example has eight outputs and can be regarded as storing an 8-bit data *byte* (the 8 outputs) for each address. Looked at in this way, the truth **code conversion** table describes *code conversion* from an input (address) code to an output (data) code.

A simple example of code conversion would be from the 3-bit natural binary code for the octal digits 0–7 to the so-called Gray cyclic code (which has the property that only one bit changes between the codes for successive octal digits, including those for 7 and 0).

The truth table is simply.

address (octal digit)	input (natural binary)			output (Gray cyclic)		
	A	B	C	P	Q	R
0	0	0	0	0	0	0
1	0	0	1	0	0	1
2	0	1	0	0	1	1
3	0	1	1	0	1	0
4	1	0	0	1	1	0
5	1	0	1	1	1	1
6	1	1	0	1	0	1
7	1	1	1	1	0	0.

Seven segment display

A rather more complicated case, which is in widespread use, is the conversion from $8:4:2:1$ BCD to the code needed to generate readable numbers on a seven segment display. This is the now familiar pattern used on digital calculators, voltmeters, clocks and so on, using seven line segments arranged as in Figure 22, which can be separately illuminated to form recognizable denary numbers. Thus, to form 1, it is only necessary to illuminate segments b and c, but not a, d, e, f and g; to form 8 it is necessary to illuminate all seven segments and so on. The truth table is therefore:

address					segment (on = 1)						
(denary digit)	(binary code)				a	b	c	d	e	f	g
0	0	0	0	0	1	1	1	1	1	1	0
1	0	0	0	1	0	1	1	0	0	0	0
2	0	0	1	0	1	1	0	1	1	0	1
3	0	0	1	1	1	1	1	1	0	0	1
4	0	1	0	0	0	1	1	0	0	1	1
5	0	1	0	1	1	0	1	1	0	1	1
6	0	1	1	0	1	0	1	1	1	1	1
7	0	1	1	1	1	1	1	0	0	0	0
8	1	0	0	0	1	1	1	1	1	1	1
9	1	0	0	1	1	1	1	1	0	1	1
no meaning	1	0	1	0							
for BCD	1	0	1	1							
	1	1	0	0							
	1	1	0	1							
	1	1	1	0							
	1	1	1	1							

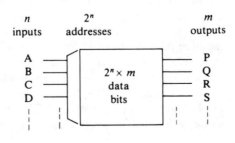

Figure 22

General properties of truth tables

You have now seen examples of truth tables with various numbers of inputs and outputs. In general, there might be n inputs and m outputs, corresponding to a combinational logic device, for example a ROM, as in Figure 23. Tabulating the various cases considered,

Figure 23

we can arrive at general expressions for the number of addresses (rows), the number of (output) data bits to be filled in, and the number of ways this can be done.

n	m	number of addresses (rows)	number of data bits	number of ways data can be arranged
1	1	$2^1 = 2$	2	$2^2 = 4$
2	1	$2^2 = 4$	4	$2^4 = 16$
3	1	$2^3 = 8$	8	$2^8 = 256$
4	1	$2^4 = 16$	16	$2^{16} = 65\,536$
3	3	$2^3 = 8$	$8 \times 3 = 24$	2^{24}
4	7	$2^4 = 16$	$16 \times 7 = 112$	2^{112}
n	m	2^n	$2^n \times m$	$2^{2^n \times m}$

One type of standard ROM has $n = 10$, $m = 8$, so $2^{10} = 1024$ (conventionally 1 K) addresses, and $8 \times 1024 = 8192 = 8$ K data locations to be filled in. This can be done in 2^{8192} ways; a vast number ($> 10^{2465}$). This emphasizes the flexibility of the ROM approach, the same device being usable as the solution to a vast number of different problems.

Electronic combinational logic

Having looked at the way combinational problems can be coded in digital form and specified by truth tables, a few typical electronic integrated circuits can be introduced that can be used to 'implement' the specification and provide practical solutions to such problems. The number of different integrated circuits to be found in manufacturers' catalogues is at first sight rather forbidding; new devices with improved properties are constantly being produced. The aim here can only be to provide a sample of the kind of information manufacturers provide so that you will be better able to look at and interpret it for yourself. As the internal structure of integrated circuits is so complex and diverse, their design will not be explained, only the effects seen at the terminals, the transfer function. The manufacturing techniques and logic circuits used combine to characterize 'compatible logic families', ranges of integrated circuits operating to the same standards and designed to be used together to build up more complicated circuits.

TTL

A widely used family is called *TTL* (transistor–transistor logic), named after the particular type of transistor circuit used to provide basic gate functions. It is to be distinguished from, for example, DTL (diode–transistor logic), ECL (emitter-coupled logic) and so on, which use different circuits and have different standards.

All of these types are based on bipolar transistor manufacturing techniques, and are all to be distinguished from *MOS* (metal-oxide-semiconductor) devices which involve different manufacturing methods and *field-effect* transistors. Again there are various MOS types (CMOS, NMOS, and so on) with different characteristics. However, some MOS devices, like the ROM to be described below, are called 'TTL compatible', which means that they will interface in some specified way with the TTL devices proper. So I will limit myself here to a brief discussion of a few TTL and compatible MOS circuits to give the flavour of the available specifications. It must be repeated that the range of devices is very wide and constantly changing, so these examples should only be regarded as illustrations.

Some SSI devices

Excerpts from early pages of the manufacturer's handbook on the '54/74 TTL family' are shown as Figure 24. This identifies a number of specfiic devices which include four 2-input gates, two 4-input gates or six inverters on one chip (thus, 'small-scale integration', SSI) and shows how they are connected to the pins of a standard 14-pin IC package. Each 2-input gate has three terminals, so 12 pins are needed for the gate connections, leaving two to connect to the power supply, V_{cc} and GND. The 4-input gates have five terminals each, so two pins are left unconnected. The inverters have two terminals each, so six of them (confusingly called hex!) use all available pins. The first group is labelled 00, meaning that the basic type number is either 5400 or 7400, with various additional letters for the devices listed, implying detailed differences in specification or packaging. They each contain four NAND gates, each with the property: (output) $\mathbf{Y} = \mathbf{AB}$, using 'positive logic', with 'high voltage' representing 1, 'low voltage' 0, just as in all our previous discussion. A practical device can only produce high and low *voltage* levels H and L; the interpretation of these as *logic* states 1 and 0 is a convention. The second group shown, 02, provides four 2-input NOR gates, the third group, 04, six ('hex') inverters, the fourth, 20, two 4-input NAND gates, and so on. There are many more types involving gates with more inputs and various forms of interconnection.

Returning to the 00 group, its electrical characteristics are reproduced here as Figure 25. There is much more information here than you need to try to take in now, but a few points are introduced. First the supply voltage V_{cc}, which is nominally 5 V, with a small range around it for all types. Next the operating temperature, which can cover a much wider range for the 54 family than for the 74 family, 54 being originally designed for

54/74 FAMILIES OF COMPATIBLE TTL CIRCUITS

PIN ASSIGNMENTS (TOP VIEWS)

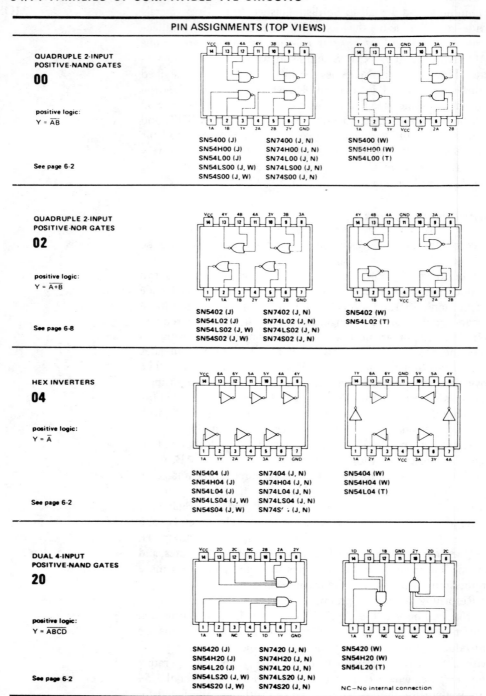

QUADRUPLE 2-INPUT POSITIVE-NAND GATES

00

positive logic:
$Y = \overline{AB}$

See page 6-2

SN5400 (J)   SN7400 (J, N)   SN5400 (W)
SN54H00 (J)   SN74H00 (J, N)   SN54H00 (W)
SN54L00 (J)   SN74L00 (J, N)   SN54L00 (T)
SN54LS00 (J, W) SN74LS00 (J, N)
SN54S00 (J, W)   SN74S00 (J, N)

QUADRUPLE 2-INPUT POSITIVE-NOR GATES

02

positive logic:
$Y = \overline{A+B}$

See page 6-8

SN5402 (J)   SN7402 (J, N)   SN5402 (W)
SN54L02 (J)   SN74L02 (J, N)   SN54L02 (T)
SN54LS02 (J, W) SN74LS02 (J, N)
SN54S02 (J, W)   SN74S02 (J, N)

HEX INVERTERS

04

positive logic:
$Y = \overline{A}$

See page 6-2

SN5404 (J)   SN7404 (J, N)   SN5404 (W)
SN54H04 (J)   SN74H04 (J, N)   SN54H04 (W)
SN54L04 (J)   SN74L04 (J, N)   SN54L04 (T)
SN54LS04 (J, W) SN74LS04 (J, N)
SN54S04 (J, W)   SN74S' . (J, N)

DUAL 4-INPUT POSITIVE-NAND GATES

20

positive logic:
$Y = \overline{ABCD}$

See page 6-2

SN5420 (J)   SN7420 (J, N)   SN5420 (W)
SN54H20 (J)   SN74H20 (J, N)   SN54H20 (W)
SN54L20 (J)   SN74L20 (J, N)   SN54L20 (T)
SN54LS20 (J, W) SN74LS20 (J, N)
SN54S20 (J, W)   SN74S20 (J, N)   NC—No internal connection

Figure 24

recommended operating conditions

		SERIES 54 SERIES 74			SERIES 54H SERIES 74H			SERIES 54L SERIES 74L			SERIES 54LS SERIES 74LS			SERIES 54S SERIES 74S			
		'00, '04, '10, '20, '30			'H00, 'H04, 'H10, 'H20, 'H30			'L00, 'L04, 'L10, 'L20, 'L30			'LS00, 'LS04, 'LS10, 'LS20, 'LS30			'S00, 'S04, 'S10, 'S20, 'S30, 'S133			UNIT
		MIN	NOM	MAX	MIN	NOM	MAX	MIN	NOM	MAX	MIN	NOM	MAX	MIN	NOM	MAX	
Supply voltage, V_{CC}	54 Family	4.5	5	5.5	4.5	5	5.5	4.5	5	5.5	4.5	5	5.5	4.5	5	5.5	
	74 Family	4.75	5	5.25	4.75	5	5.25	4.75	5	5.25	4.75	5	5.25	4.75	5	5.25	
High-level output current, I_{OH}	54 Family			−400			−500			−100			−400			−1000	μA
	74 Family			−400			−500			−200			−400			−1000	
Low-level output current, I_{OL}	54 Family			16			20			2			4			20	mA
	74 Family			16			20			3.6			8			20	
Operating free-air temperature, T_A	54 Family	−55		125	−55		125	−55		125	−55		125	−55		125	°C
	74 Family	0		70	0		70	0		70	0		70	0		70	

electrical characteristics over recommended operating free-air temperature range (unless otherwise noted)

PARAMETER		TEST FIGURE	TEST CONDITIONS[†]		SERIES 54 SERIES 74			SERIES 54H SERIES 74H			SERIES 54L SERIES 74L			SERIES 54LS SERIES 74LS			SERIES 54S SERIES 74S			UNIT
					'00, '04, '10, '20, '30			'H00, 'H04, 'H10, 'H20, 'H30			'L00, 'L04, 'L10, 'L20, 'L30			'LS00, 'LS04, 'LS10, 'LS20, 'LS30			'S00, 'S04, 'S10, 'S20, 'S30, 'S133			
					MIN	TYP[‡]	MAX	MIN	TYP[‡]	MAX	MIN	TYP[‡]	MAX	MIN	TYP[‡]	MAX	MIN	TYP[‡]	MAX	
V_{IH}	High-level input voltage	1, 2			2			2			2			2			2			V
V_{IL}	Low-level input voltage	1, 2		54 Family			0.8			0.8			0.7			0.7			0.8	V
				74 Family			0.8			0.8			0.7			0.8			0.8	
V_{IK}	Input clamp voltage	3	V_{CC} = MIN, I_I =				−1.5			−1.5						−1.5			−1.2	V
V_{OH}	High-level output voltage	1	V_{CC} = MIN, V_{IL} = V_{IL} max, I_{OH} = MAX	54 Family	2.4	3.4		2.4	3.5		2.4	3.3		2.5	3.4		2.5	3.4		V
				74 Family	2.4	3.4		2.4	3.5		2.4	3.2		2.7	3.4		2.7	3.4		
V_{OL}	Low-level output voltage	2	V_{CC} = MIN, I_{OL} = MAX V_{IH} = 2 V	54 Family		0.2	0.4		0.2	0.4		0.15	0.3		0.25	0.4			0.5	V
				74 Family		0.2	0.4		0.2	0.4		0.2	0.4		0.25	0.5			0.5	
			I_{OL} = 4 mA	Series 74LS												0.4				
I_I	Input current at maximum input voltage	4	V_{CC} = MAX	V_I = 5.5 V			1			1			0.1						1	mA
				V_I = 7 V												0.1				
I_{IH}	High-level input current	4	V_{CC} = MAX	V_{IH} = 2.4 V			40			50			10							μA
				V_{IH} = 2.7 V												20			50	
I_{IL}	Low-level input current	5	V_{CC} = MAX	V_{IL} = 0.3 V									−0.18							mA
				V_{IL} = 0.4 V			−1.6			−2						−0.4				
				V_{IL} = 0.5 V															−2	
I_{OS}	Short-circuit output current[◆]	6	V_{CC} = MAX	54 Family	−20		−55	−40		−100	−3		−15	−20		−100	−40		−100	mA
				74 Family	−18		−55	−40		−100	−3		−15	−20		−100	−40		−100	
I_{CC}	Supply current	7	V_{CC} = MAX								See table on next page									mA

[†] For conditions shown as MIN or MAX, use the appropriate value specified under recommended operating conditions.
[‡] All typical values are at V_{CC} = 5 V, T_A = 25°C.
[§] I_I = −12 mA for SN54'/SN74', −8 mA for SN54H'/SN74H', and −18 mA for SN54LS'/SN74LS' and SN54S'/SN74S'.
[◆] Not more than one output should be shorted at a time, and for SN54H'/SN74H', SN54LS'/SN74LS', and SN54S'/SN74S', duration of short-circuit should not exceed 1 second.

supply current[¶]

TYPE	I_{CCH} (mA) Total with outputs high		I_{CCL} (mA) Total with outputs low		I_{CC} (mA) Average per gate (50% duty cycle)
	TYP	MAX	TYP	MAX	TYP
'00	4	8	12	22	2
'04	6	12	18	33	2
'10	3	6	9	16.5	2
'20	2	4	6	11	2
'30	1	2	3	6	2
'H00	10	16.8	26	40	4.5
'H04	16	26	40	58	4.5
'H10	7.5	12.6	19.5	30	4.5
'H20	5	8.4	13	20	4.5
'H30	2.5	4.2	6.5	10	4.5
'L00	0.44	0.8	1.16	2.04	0.20
'L04	0.66	1.2	1.74	3.06	0.20
'L10	0.33	0.6	0.87	1.53	0.20
'L20	0.22	0.4	0.58	1.02	0.20
SN54L30	0.11	0.33	0.29	0.51	0.20
SN74L30	0.11	0.2	0.29	0.51	0.20
'LS00	0.8	1.6	2.4	4.4	0.4
'LS04	1.2	2.4	3.6	6.6	0.4
'LS10	0.6	1.2	1.8	3.3	0.4
'LS20	0.4	0.8	1.2	2.2	0.4
'LS30	0.35	0.5	0.6	1.1	0.48
'S00	10	16	20	36	3.75
'S04	15	24	30	54	3.75
'S10	7.5	12	15	27	3.75
'S20	5	8	10	18	3.75
'S30	3	5	5.5	10	4.25
'S133	3	5	5.5	10	4.25

[¶] Maximum values of I_{CC} are over the recommended operating ranges of V_{CC} and T_A; typical values are at V_{CC} = 5 V, T_A = 25°C.

switching characteristics at V_{CC} = 5 V, T_A = 25°C

TYPE	TEST CONDITIONS[#]	t_{PLH} (ns) Propagation delay time, low-to-high-level output			t_{PHL} (ns) Propagation delay time, high-to-low-level output		
		MIN	TYP	MAX	MIN	TYP	MAX
'00, '10	C_L = 15 pF, R_L = 400 Ω		11	22		7	15
'04, '20			12	22		8	15
'30			13	22		8	15
'H00	C_L = 25 pF, R_L = 280 Ω		5.9	10		6.2	10
'H04			6	10		6.5	10
'H10			5.9	10		6.3	10
'H20			6	10		7	10
'H30			6.8	10		8.9	12
'L00, 'L04, 'L10, L20	C_L = 50 pF, R_L = 4 kΩ		35	60		31	60
'L30			35	60		70	100
'LS00, 'LS04 'LS10, 'LS20	C_L = 15 pF, R_L = 2 kΩ		9	15		10	15
'LS30			8	15		13	20
'S00, 'S04	C_L = 15 pF, R_L = 280 Ω		3	4.5		3	5
'S10, 'S20	C_L = 50 pF, R_L = 280 Ω		4.5			5	
'S30, 'S133	C_L = 15 pF, R_L = 280 Ω		4	6		4.5	7
	C_L = 50 pF, R_L = 280 Ω		5.5			6.5	

[#] Load circuits and voltage waveforms are shown on pages 3-10 and 3-11.

schematics (each gate)

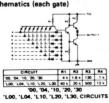

CIRCUIT	R1	R2	R3	R4
'00, '04, '10, '20, '30	4 k	1.6 k	130	1 k
'L00, 'L04, 'L10, 'L20, 'L30	40 k	20 k	500	12 k

'00, '04, '10, '20, '30
'L00, 'L04, 'L10, 'L20, 'L30, CIRCUITS

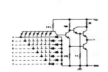

'H00, 'H04, 'H10, 'H20, 'H30 CIRCUITS

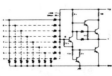

'LS00, 'LS04, 'LS10, 'LS20, 'LS30 CIRCUITS

'S00, 'S04, 'S10, 'S20, 'S30, 'S133 CIRCUITS

Figure 25

military use. Then the voltage levels representing logic states, which, with V_{cc}, are the distinctive characteristics of TTL that 'TTL compatible' devices have to match. A 'high' logic *input* V_{IH} must be greater than 2 V, a 'low' V_{IL} less than 0.8 V; logic *outputs* produced will then be high, V_{OH}, greater than 2.4 V, and low, V_{OL}, less than 0.4 V. These differences of 0.4 V between specified extreme input and output levels provide a 'noise margin', so that the output of one circuit can be passed on to the input of the next without error even if random noise disturbs the voltage levels to that extent.

Some MSI devices

BCD tp 7-segment decoders/drivers

In viewing representative 'medium scale' (MSI) circuits, Figure 26 shows some of the catalogue information about *BCD to 7-segment decoders/drivers*. There are various types, some with additional features that need not be discussed here, but all with a 'function table' similar to that at the bottom of the figure. This is like the truth table you have seen before except that

(*i*): It is expressed in terms of 'high' and 'low' voltages, H and L, to avoid any confusion over logic conventions;

(*ii*): it describes rather meaningless symbols for addresses 10–15, which of course should not arise with BCD inputs; and

(*iii*): it has an additional 'blanking' input, which as well as providing a blank display when required can also be switched rapidly from H to L to provide an 'intensity modulation capability', that is a control of the apparent brightness of the display.

An LSI device

mask-programmed ROM

programmable ROM

All the devices described so far, of varying complexity, have fixed built-in truth tables. However, there is a range of general purpose combinational devices, the read-only memories or ROMs, which can be 'programmed' with any required truth table, limited only to a certain size for each particular type. In *'mask-programmed''* ROMs the decision about the required truth table has to be made at the time of ordering; the manufacturer designs appropriate 'masks' (as part of the photographic processing) for the particular requirement. This is a relatively expensive procedure, and is only appropriate when a design is settled and large numbers of identical ROMs are needed so that the unit cost is then low. *'Programmable' ROM's* are more appropriate for the development of prototype systems, as they are supplied with an effectively blank truth table which can be converted into the required one by applying pulses in

'46A, '47A, 'L46, 'L47, 'LS47 feature	'48, 'LS48 feature	'49, 'LS49 feature
• Open-Collector Outputs Drive Indicators Directly	• Internal Pull-Ups Eliminate Need for External Resistors	• Open-Collector Outputs
• Lamp-Test Provision	• Lamp-Test Provision	• Blanking Input
• Leading/Trailing Zero Suppression	• Leading/Trailing Zero Suppression	

• All Circuit Types Feature Lamp Intensity Modulation Capability

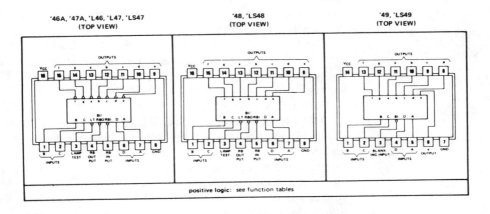

'46A, '47A, 'L46, 'L47, 'LS47 (TOP VIEW)   '48, 'LS48 (TOP VIEW)   '49, 'LS49 (TOP VIEW)

positive logic: see function tables

SEGMENT IDENTIFICATION

NUMERICAL DESIGNATIONS AND RESULTANT DISPLAYS

'49, 'LS49 FUNCTION TABLE

DECIMAL OR FUNCTION	INPUTS					OUTPUTS							NOTE
	D	C	B	A	BI	a	b	c	d	e	f	g	
0	L	L	L	L	H	H	H	H'	H	H	H	L	
1	L	L	L	H	H	L	H	H	L	L	L	L	
2	L	L	H	L	H	H	H	L	H	H	L	H	
3	L	L	H	H	H	H	H	H	H	L	L	H	
4	L	H	L	L	H	L	H	H	L	L	H	H	
5	L	H	L	H	H	H	L	H	H	L	H	H	
6	L	H	H	L	H	L	L	H	H	H	H	H	
7	L	H	H	H	H	H	H	H	L	L	L	L	1
8	H	L	L	L	H	H	H	H	H	H	H	H	
9	H	L	L	H	H	H	H	H	L	L	H	H	
10	H	L	H	L	H	L	L	L	H	H	L	H	
11	H	L	H	H	H	L	L	H	H	L	L	H	
12	H	H	L	L	H	L	H	L	L	L	H	H	
13	H	H	L	H	H	H	L	L	H	L	H	H	
14	H	H	H	L	H	L	L	L	H	H	H	H	
15	H	H	H	H	H	L	L	L	L	L	L	L	
BI	X	X	X	X	L	L	L	L	L	L	L	L	2

H = high level, L = low level, X = irrelevant

NOTES: 1. The blanking input (BI) must be open or held at a high logic level when output functions 0 through 15 are desired.
2. When a low logic level is applied directly to the blanking input (BI), all segment outputs are low regardless of the level of any other input.

Figure 26

some specified way to the pins. The detailed specifications vary; the fairly detailed information of Figure 27 focuses on one particular type of 'EPROM', which as well as being 'electrically programmable' is also 'u.v. erasable'. This means that the required truth table can be programmed into the device by applying suitable electrical signals, but can also be erased if it is wrong, or required to be changed, by exposure to ultraviolet light; the package has a transparent window so that the light can fall on the chip. Although the specification is very detailed, here are some key ideas. The first is that, although this is an MOS device, it has been designed to operate on the same $+5$ V power supply and with the same H and L voltage levels to represent the two logic states as a TTL device. Like any ROM it has 'address' inputs, in this case 10 of them, A_0-A_9 (so $2^{10} = 1024 = 1$ K addresses) and 'data' outputs, in this case 8 of them, D_0-D_7 (so one byte of data at each address). For programming it requires a second power supply input (V_{pp}) to be raised from $+5$ V to $+25$ V, but otherwise only suitable TTL voltage levels at three further pins (A_R, PD/PGM and $\overline{CS}$).

The various options are summarized in the MODE SELECTION table. A_R must always be L; if both the other control inputs PD/PGM and $\overline{CS}$ are also L this is the normal 'read' condition, as for any ROM, giving at the outputs whatever data has been programmed in at each address. (Note the use of an inversion bar on $\overline{CS}$ to mean that the chip selection is done by applying a *low* voltage.) If PD/PGM is raised to H this puts the chip into the 'power down' mode, with reduced power dissipation. In this case, or if $\overline{CS}$ is taken high, there is a high impedance at the outputs which effectively disconnects them. (Such devices are said to have three output states. More will be said about these 'tristate' outputs, with high impedance as well as high and low voltage states, in chapter 5.) Programming requires $V_{pp} = +25$ V (and $C_{cc} = +5$ V as usual) and $\overline{CS}$ to be H (to deselect the output). For a particular address (applied to the address pins) the required data is then applied to the output pins (8 bits at a time) and a program pulse applied to take PD/PGM from L to H for 50 ms. This is then repeated with the appropriate data for each address in turn. If $\overline{CS}$ is brought to L while PD/PGM is L this *verifies* the program; that is, produces the stored data at the output pins. If PD/PGM is kept L while $\overline{CS}$ is H this *inhibits* the program, that is provides a high impedance at the output. This means that two 2758s can have all corresponding pins, except PD/PGM, connected together, and only the one to which a program pulse is applied will store the data pins at the address on the address pins. Although an EPROM of this kind can be programmed and re-programmed relatively easily, it is still intended that in normal use, like any ROM, it will be a fixed memory, to be 'read' only. More will be said about the relationship between ROMs and another type of memory device, a RAM (or better 'read-write memory') which is similar in its storage of data which can be read at particular addresses, but can also have data 'written' into the address during normal operation.

LOWPOWER, +5V 1Kx8 UV EPROM

- **Single +5V Power Supply**

- **Simple Programming Requirements**
  Single Location Programming
  Programs With One 50ms Pulse

- **Low Power Dissipation**
  525mW Max Active Power
  132mW Max Standby Power

- **Fast Access Time: 450ns Max In**
  Active and Standby Power Mode

- **Inputs and Outputs TTL Compatible**
  During Read and Program

- **Three-State Outputs For or-Ties**

The Intel® 2758 is a 8192-bit ultraviolet erasable and electrically programmable read-only-memory (EPROM). The 2758 operates from a single 5V power supply, has a static power-down mode, and features fast, single address location programming. It makes designing with EPROMs faster, easier and more economical. The total programming time for all 8192 bits is 50 seconds.

The 2758 has a static power-down mode which reduces the power dissipation without increasing access time. The maximum active power dissipation is 525 mW, while the maximum standby power dissipation is only 132 mW, a 75% savings. Power-down is achieved by applying a TTL-high signal to the PD/PGM input.

A 2758 system may be designed for total upwards compatibility with Intel's 16K 2716 EPROM (see Applications Note 30). The 2758 maintains the simplest and fastest method yet devised for programming EPROMs — single pulse TTL-level programming. There is no need for high voltage pulsing because all programming controls are handled by TTL signals. Now it is possible to program on-board, in the system, in the field. Program any location at any time — either individually, sequentially, or at random, with the single address location programming.

PIN CONFIGURATION

```
         ┌───┐
A7  ☐ 1  └─┘ 24 ☐ VCC
A6  ☐ 2      23 ☐ A8
A5  ☐ 3      22 ☐ A9
A4  ☐ 4      21 ☐ VPP
A3  ☐ 5      20 ☐ CS̄
A2  ☐ 6      19 ☐ AR
A1  ☐ 7      18 ☐ PD/PGM
A0  ☐ 8      17 ☐ O7
O0  ☐ 9      16 ☐ O6
O1  ☐ 10     15 ☐ O5
O2  ☐ 11     14 ☐ O4
GND ☐ 12     13 ☐ O3
```

MODE SELECTION

MODE \ PINS	PD/PGM (18)	A_R (19)	$\overline{CS}$ (20)	V_{PP} (21)	V_{CC} (24)	OUTPUTS (9-11, 13-17)
Read	V_{IL}	V_{IL}	V_{IL}	+5	+5	D_{OUT}
Deselect	Don't Care	V_{IL}	V_{IH}	+5	+5	High Z
Power Down	V_{IH}	V_{IL}	Don't Care	+5	+5	High Z
Program	Pulsed V_{IL} to V_{IH}	V_{IL}	V_{IH}	+25	+5	D_{IN}
Program Verify	V_{IL}	V_{IL}	V_{IL}	+25	+5	D_{OUT}
Program Inhibit	V_{IL}	V_{IL}	V_{IH}	+25	+5	High Z

PIN NAMES

A_0-A_9	ADDRESSES
PD/PGM	POWER DOWN/PROGRAM
$\overline{CS}$	CHIP SELECT
O_0-O_7	OUTPUTS
A_R	SELECT REFERENCE INPUT LEVEL

BLOCK DIAGRAM

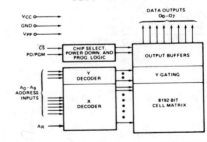

August 1977

Figure 27

READ OPERATION

D.C. and Operating Characteristics

$T_A = 0°C$ to $70°C$, $V_{CC}^{[1,2]} = +5V \pm 5\%$, $V_{PP}^{[2]} = V_{CC} \pm 0.6V^{[3]}$

Symbol	Parameter	Limits			Unit	Conditions
		Min.	Typ.[4]	Max.		
I_{LI}	Input Load Current			10	μA	$V_{IN} = 5.25V$
I_{LO}	Output Leakage Current			10	μA	$V_{OUT} = 5.25V$
I_{PP1} [2]	V_{PP} Current			5	mA	$V_{PP} = 5.85V$
I_{CC1} [2]	V_{CC} Current (Standby)		10	25	mA	PD/PGM = V_{IH}, $\overline{CS}$ = V_{IL}
I_{CC2} [2]	V_{CC} Current (Active)		57	100	mA	$\overline{CS}$ = PD/PGM = V_{IL}
A_R [5]	Select Reference Input Level	-0.1		0.8	V	$I_{IN} = 10 \mu A$
V_{IL}	Input Low Voltage	-0.1		0.8	V	
V_{IH}	Input High Voltage	2.2		V_{CC} + 1	V	
V_{OL}	Output Low Voltage			0.45	V	$I_{OL} = 2.1$ mA
V_{OH}	Output High Voltage	2.4			V	$I_{OH} = -400 \mu A$

Typical Characteristics

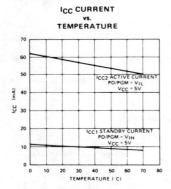

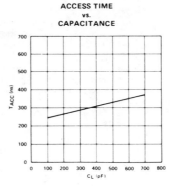

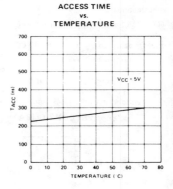

A.C. Characteristics

$T_A = 0°C$ to $70°C$, $V_{CC}^{[1]} = +5V \pm 5\%$, $V_{PP}^{[2]} = V_{CC} \pm 0.6V^{[3]}$

Symbol	Parameter	Limits			Unit	Test Conditions
		Min.	Typ.[4]	Max.		
t_{ACC1}	Address to Output Delay		250	450	ns	PD/PGM = $\overline{CS}$ = V_{IL}
t_{ACC2} [7]	PD/PGM to Output Delay		280	450	ns	$\overline{CS}$ = V_{IL}
t_{CO}	Chip Select to Output Delay			120	ns	PD/PGM = V_{IL}
t_{PF}	PD/PGM to Output Float	0		100	ns	$\overline{CS}$ = V_{IL}
t_{DF}	Chip Deselect to Output Float	0		100	ns	PD/PGM = V_{IL}
t_{OH}	Address to Output Hold	0			ns	PD/PGM = $\overline{CS}$ = V_{IL}

Capacitance [6] $T_A = 25°C$, $f = 1$ MHz

Symbol	Parameter	Typ.	Max.	Unit	Conditions
C_{IN}	Input Capacitance	4	6	pF	$V_{IN} = 0V$
C_{OUT}	Output Capacitance	8	12	pF	$V_{OUT} = 0V$

NOTE: Please refer to page 2 for notes.

A.C. Test Conditions:

Output Load: 1 TTL gate and $C_L = 100$ pF
Input Rise and Fall Times: $\leqslant 20$ ns
Input Pulse Levels: 0.8V to 2.2V
Timing Measurement Reference Level:
   Inputs   1V and 2V
   Outputs   0.8V and 2V

Figure 27 (continued)

WAVEFORMS

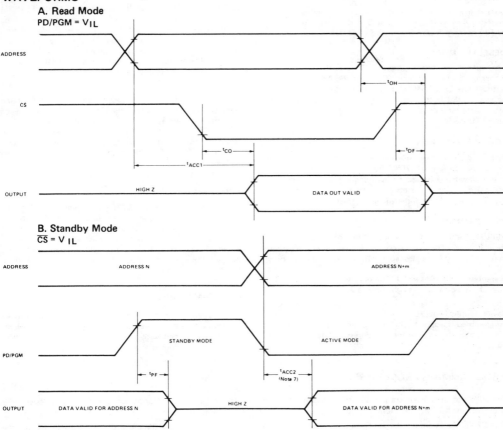

A. Read Mode
PD/PGM = V_{IL}

ADDRESS

CS

t_{OH}

t_{CO}

t_{ACC1}

t_{DF}

OUTPUT — HIGH Z — DATA OUT VALID

B. Standby Mode
$\overline{CS} = V_{IL}$

ADDRESS — ADDRESS N — ADDRESS N+m

PD/PGM — STANDBY MODE — ACTIVE MODE

t_{PF}

t_{ACC2} (Note 7)

OUTPUT — DATA VALID FOR ADDRESS N — HIGH Z — DATA VALID FOR ADDRESS N+m

ERASURE CHARACTERISTICS

The erasure characteristics of the 2758 are such that erasure begins to occur when exposed to light with wavelengths shorter than approximately 4000 Angstroms (Å). It should be noted that sunlight and certain types of fluorescent lamps have wavelengths in the 3000—4000Å range. Data show that constant exposure to room level fluorescent lighting could erase the typical 2758 in approximately 3 years, while it would take approximately 1 week to cause erasure when exposed to direct sunlight. If the 2758 is to be exposed to these types of lighting conditions for extended periods of time, opaque labels are available from Intel which should be placed over the 2758 window to prevent unintentional erasure.

The recommended erasure procedure for the 2758 is exposure to shortwave ultraviolet light which has a wavelength of 2537 Angstroms (Å). The integrated dose (i.e., UV intensity $\times$ exposure time) for erasure should be a minimum of 15 W-sec/cm^2. The erasure time with this dosage is approximately 15 to 20 minutes using an ultraviolet lamp with a 12,000 μW/cm^2 power rating. The 2758 should be placed within 1 inch of the lamp tubes during erasure. Some lamps have a filter on their tubes which should be removed before erasure.

DEVICE OPERATION

The six modes of operation of the 2758 are listed in Table I. It should be noted that all inputs for the six modes are at TTL levels. The power supplies required are a +5V V_{CC} and a V_{PP}. The V_{PP} power supply must be at 25V during the three programming modes, and must be at 5V in the other three modes. In all operational modes, A_R must be at V_{IL} (except for the 2758 S1865 which has A_R at V_{IH}).

TABLE I. MODE SELECTION

MODE \ PINS	PD/PGM (18)	A_R (19)	$\overline{CS}$ (20)	V_{PP} (21)	V_{CC} (24)	OUTPUTS (9-11, 13-17)
Read	V_{IL}	V_{IL}	V_{IL}	+5	+5	D_{OUT}
Deselect	Don't Care	V_{IL}	V_{IH}	+5	+5	High Z
Power Down	V_{IH}	V_{IL}	Don't Care	+5	+5	High Z
Program	Pulsed V_{IL} to V_{IH}	V_{IL}	V_{IH}	+25	+5	D_{IN}
Program Verify	V_{IL}	V_{IL}	V_{IL}	+25	+5	D_{OUT}
Program Inhibit	V_{IL}	V_{IL}	V_{IH}	+25	+5	High Z

Figure 27 (continued)

READ MODE

Data is available at the outputs in the read mode. Data is available 450 ns (t_{ACC}) from stable addresses with $\overline{CS}$ low or 120 ns (t_{CO}) from $\overline{CS}$ with addresses stable.

DESELECT MODE

The outputs of two or more 2758 may be OR-tied together on the same data bus. Only one 2758 should have its outputs selected ($\overline{CS}$ low) to prevent data bus contention between 2758 in this configuration. The outputs of the other 2758 should be deselected with the $\overline{CS}$ input at a high TTL level.

POWER DOWN MODE

The 2758 has a power down mode which reduces the active power dissipation by 75%, from 525 mW to 132 mW. Power down is achieved by applying a TTL high signal to the PD/PGM input. In power down the outputs are in a high impedance state, independent of the $\overline{CS}$ input.

PROGRAMMING

Initially, and after each erasure, all bits of the 2758 are in the "1" state. Data is introduced by selectively programming "0's" into the desired bit locations. Although only "0's" will be programmed, both "1's" and "0's" can be presented in the data word. The only way to change a "0" to a "1" is by ultraviolet light erasure.

The 2758 is in the programming mode when the V_{PP} power supply is at 25V and $\overline{CS}$ is at V_{IH}. The data to be programmed is applied 8 bits in parallel to the data output pins. The levels required for the address and data inputs are TTL.

When the addresses and data are stable, a 50 msec, active high, TTL program pulse is applied to the PD/PGM input. A program pulse must be applied at each address location to be programmed. You can program any location at any time — either individually, sequentially, or at random. The program pulse has a maximum width of 55 msec. The 2758 must not be programmed with a DC signal applied to the PD/PGM input.

Programming of multiple 2758s in parallel with the same data can be easily accomplished due to the simplicity of the programming requirements. Like inputs of the parallelled 2758s may be connected together when they are programmed with the same data. A high level TTL pulse applied to the PD/PGM input programs the paralleled 2758s.

PROGRAM INHIBIT

Programming of multiple 2758s in parallel with different data is also easily accomplished. Except for PD/PGM, all like inputs (including $\overline{CS}$) of the parallel 2758s may be common. A TTL level program pulse applied to a 2758's PD/PGM input with V_{PP} at 25V will program that 2758. A low level PD/PGM input inhibits the other 2758 from being programmed.

PROGRAM VERIFY

A verify should be performed on the programmed bits to determine that they were correctly programmed. The verify may be performed with V_{PP} at 25V. Except during programming and program verify, V_{PP} must be at 5V.

PROGRAMMING PROCEDURES

Initially, and after each erasure, all 8,192 bits of the 2758 are in the "1" state. Information is introduced by selectively programming "0" into the desired bit locations. A programmed "0" can only be changed to a "1" by UV erasure.

The 2758 is programmed by applying a 50 ms, TTL programming pulse to the PD/PGM pin with the $\overline{CS}$ input high and the V_{PP} supply at 25V ±1V. Any location may be programmed at any time — either individually, sequentially, or randomly. The programming time for a single bit is only 50 ms and for all 8,192 bits is approximately 50 sec. The detailed programming specifications and timing waveforms are given in the following tables and figures.

2758 PROGRAM CHARACTERISTICS [1]

$T_A = 25°C ±5°C$, V_{CC}[2] = 5V ±5%, V_{PP}[2,3] = 25V ±1V

D.C. Programming Characteristics

Symbol	Parameter	Min.	Typ.	Max.	Units	Test Conditions
I_{LI}	Input Current (for Any Input)			10	μA	V_{IN} = 5.25V/0.45
I_{PP1}	V_{PP} Supply Current			5	mA	PD/PGM = V_{IL}
I_{PP2}	V_{PP} Supply Current During Programming Pulse			30	mA	PD/PGM = V_{IH}
I_{CC}	V_{CC} Supply Current			100	mA	
V_{IL}	Input Low Level	−0.1		0.8	V	
V_{IH}	Input High Level	2.2		V_{CC}+1	V	
A_R	Select Reference Input Level	−0.1		0.8	V	I_{IN} = 10 μA

Figure 27 (continued)

A.C. Programming Characteristics

Symbol	Parameter	Min.	Typ.	Max.	Units	
t_{AS}	Address Setup Time	2			μs	
t_{CSS}	$\overline{CS}$ Setup Time	2			μs	
t_{DS}	Data Setup Time	2			μs	
t_{AH}	Address Hold Time	2			μs	
t_{CSH}	$\overline{CS}$ Hold Time	2			μs	
t_{DH}	Data Hold Time	2			μs	
t_{DF}	Chip Deselect to Output Float Delay	0		120	ns	PD/PGM = V_{IL}
t_{CO}	Chip Select to Output Delay			120	ns	PD/PGM = V_{IL}
t_{PW}	Program Pulse Width	45	50	55	ms	
t_{PRT}	Program Pulse Rise Time	5			ns	
t_{PFT}	Program Pulse Fall Time	5			ns	

A.C. Conditions of Test:

V_{CC} . . . . 5V ±5%   Input Pulse Levels . . . . .0.8V to 2.2V
V_{PP}. . . . . 25V ±1V   Input Timing Reference Level . . . . 1V and 2V
Input Rise and Fall Times (10% to 90%) . . . . 20 ns   Output Timing Reference Level . . . . 0.8V and 2V

PROGRAMMING WAVEFORMS

V_{PP} = 25V ±1V, V_{CC} = 5V ±5%

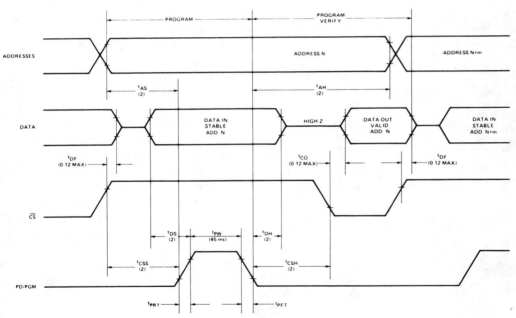

NOTE ALL TIMES SHOWN IN PARENTHESES ARE MINIMUM TIMES AND ARE μSEC UNLESS OTHERWISE NOTED

Figure 27 (continued)

Propagation delay

Combinational logic is characterized by the principle that the outputs at any time should depend only on the inputs present at that same time. That is, there is a *static* relationship between a set of inputs and a set of outputs, described by the .truth table. However, this does not mean that inputs and outputs cannot change. If one or more inputs change, the outputs must change accordingly, to the values at the new address in the truth table. This cannot happen instantaneously. Indeed, changes of voltage at *one* terminal cannot happen instantaneously, although with modern circuits they might only take a nanosecond or so. The time between some suitably defined point on the rising or falling voltage edge at an input terminal to the corresponding point on the voltage edge at an output terminal is called the *'propagation delay'* t_{prop}.

propagation delay

It is conventionally represented as in Figure 28, with a crossover to represent the possibility of voltage changes either from H to L or from L to H. Maximum and minimum values of t_{prop} should be quoted in device specifications; sometimes the values for L to H changes differ from those for H to L. As you have seen they can range from a few to a few hundred nanoseconds, for different devices. In the case of the 2758 EPROM the basic propagation delay (from 'valid address' to 'valid output') is described as the access time t_{acc}; there are also different delay times corresponding to the operation of the inputs PD/PGM and $\overline{CS}$, as Figure 27 shows. If several combinational devices are connected to form a more complex circuit, there may be more than one path, involving different devices, by which voltage changes at one point can cause effects at another. To take a simple example, if a voltage change from L to H is applied to the circuit of Figure 29, the NAND-gate output will be initially H (inputs L and H) and finally H (inputs H and L). But there will be a brief period during the propagation delay of the inverter when the NAND-gate inputs are both H, giving a corresponding brief L at the output. Complex combinational devices may well have internal alternative paths of this kind; the only way to ensure no confusion after an input change is to neglect any effects at the output until after the maximum propagation delay of the device has elapsed.

Furthermore, in a device with multiple inputs and outputs, changes cannot take place simultaneously on all inputs or on all outputs. There will be a certain 'settling' time while one required set of inputs changes to another required set of inputs (becomes 'valid' again). There will then be a delay determined by the *maximum* internal propagation delay before the outputs settle to

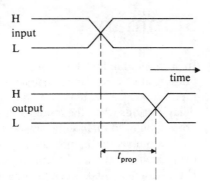

Figure 28

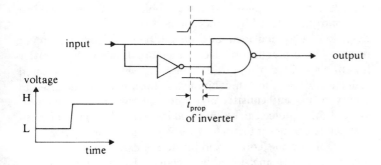
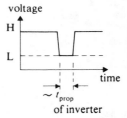

Figure 29

the new required set of valid output values. On the other hand, when inputs start to change, the delay before outputs start changing (become 'invalid') will be determined by the *minimum* internal propagation delay. All this has a bearing on the timing of sequential operations, and the minimum time which must elapse between changes, to be discussed later.

47

Summary

In a combinational logic problem or device the one or more outputs at a particular time depend on the combined effect of the input conditions at that same time (except for brief 'propagation delay' periods when the input has just changed and the output has not yet followed it). In some practical problems the input conditions and output requirements can naturally be expressed in binary or two-state form (ON/OFF conditions, YES/NO answers, TRUE/FALSE statements). A combination of n two-state input conditions defines 2^n possible input *states*. If there are m two-state output requirements, the conditions of the problem can be specified in a $2^n \times m$ truth table, with 2^n rows and m output columns. The n bits describing a particular input state can be described as an input *codeword*. In a practical read-only memory device with a transfer function specified by the truth table the input code word is often called an *address*, the output bits *data*. If the input is not naturally in binary form, but consists of a choice between individuals or symbols, the choice can be specified by a suitable binary code. An n-bit code can describe up to 2^n distinguishable inputs. A $2^n \times m$ truth table or the corresponding device can be seen as a *code converter* between an n-bit address code and an m-bit data code.

Truth tables of varying complexity can be built up from word descriptions of combinational problems. Certain simple truth tables have named functions: NOT, AND, OR, NAND, NOR, XOR; with corresponding Boolean notation $\overline{A}$, AB, $A + B$, $\overline{AB}$, $\overline{A + B}$, $A \oplus B$, which can be systematically extended to more than two inputs (see p. 49).

Each output column of an n-input truth table can be filled in one of of 2^{2^n} ways, a number which increases very rapidly with n. These will contain many cases which do not depend on all n inputs, and many more which do. In general, they can always be expressed as *'sums of minterms'* of the form $ABCD + ABC\overline{D} + AB\overline{C}D + \ldots$

Some practical electronic combinational devices have fixed truth tables, corresponding to simple logic functions (gates) or containing SSI, MSI, LSI combinations of gates to satisfy commonly encountered requirements (7-segment decoders, and so on). ROMs have a specified number of inputs and outputs, but the truth table linking them can be chosen arbitrarily. They may be 'mask-programmed' by the manufacturers (ROM) or 'programmable' by the user in some way (PROM); some can be erased and re-programmed (EPROM). When inputs are changed there is a *propagation delay* before the corresponding output change takes place.

NOT	A	$\overline{A}$
	0	1
	1	0

AND	A	B	$A.B$
	0	0	0
	0	1	0
	1	0	0
	1	1	1

OR	A	B	$A + \overline{B}$
	0	0	0
	0	1	1
	1	0	1
	1	1	1

NAND	A	B	$\overline{AB}$
	0	0	1
	0	1	1
	1	0	1
	1	1	0

NOR	A	B	$\overline{A + B}$
	0	0	1
	0	1	0
	1	0	0
	1	1	0

XOR	A	B	$A \oplus B$
	0	0	0
	0	1	1
	1	0	1
	1	1	0

49

P = A B C D | P

A	B	C	D	P
0	0	0	0	
0	0	0	1	
0	0	1	0	
0	0	1	1	
0	1	0	0	
0	1	0	1	
0	1	1	0	
0	1	1	1	
1	0	0	0	
1	0	0	1	
1	0	1	0	
1	0	1	1	
1	1	0	0	
1	1	0	1	
1	1	1	0	
1	1	1	1	

P =

Problems for Chapter 2

1. A domestic burglar alarm requires four inputs:

   A a door has been opened

   B a window has been opened

   C the alarm has been turned on

   D test the alarm bell

   and one output

   P sound the alarm bell.

   Draw up the truth table that defines the relationship between the inputs and the outputs.

2. Express the denary number 431 in natural binary, 8:4:2:1 BCD, octal and hex.

3. What sequence of 7 bit ASCII binary codes would be required to code the word EAT?

4. Derive Boolean expressions for the two output columns of the truth table below:

A	B	P_1	P_2
0	0	0	1
0	1	1	1
1	0	0	0
1	1	0	1.

5. Derive alternative expressions for:

   $$\bar{A} \oplus \bar{B} \text{ and } \bar{A} \oplus B.$$

6. A voltmeter has a cut-out and warning light which operate (**P** = 1) is an a.c./d.c. switch (**A**) is not set to correspond to the actual input (**B**), or if the input exceeds the rating of the instrument (**C**). Write down the truth table and a Boolean expression for **P**.

7. Complete this truth table for a full adder:

C_p	A	B	C_u	S
0	0	0		
0	0	1		
0	1	0		
0	1	1		
1	0	0		
1	0	1		
1	1	0		
1	1	1		

8. Although designed primarily to display numbers, the seven segment display can also produce some recognizable letters. Using the six unused output codes of the truth table for the seven segment display, complete the 'segment' columns to produce a display of

to represent the letters A B C D E F. Why would this together with the ten BCD digits not produce a satisfactory hex display?

9. Derive a Boolean expression for the logic gate circuit shown below

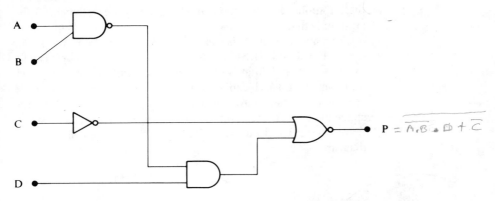

$P = \overline{A.B.D + \overline{C}}$

10. What are the maximum and typical propagation delay times for the following circuit, when the input switches from high to low? The inverters are TTL 7404 type L

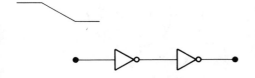

Appendix to Chapter 2

Venn diagrams

The ideas of binary variables, coding, minterms and standard logic functions may be illustrated graphically by means of Venn diagrams. This brief account is included as an appendix and you may find it provides a useful additional way of looking at the ideas. A binary variable can correspond to what is called a *'dichotomous' classification*; that is, a classification of a set of individuals or objects into just *two* subsets. Consider for example a large, but limited set of people; say all OU students in a particular year. They could be represented in a diagram by a large number of dots with a closed curve around them, separating them off from the rest of the population, as in Figure 30. The students could be classified in various ways: 'dichotomous' classifications will be those that divide them clearly into two subsets only, without uncertainties; and which can therefore be expressed in binary form. One dichotomous classification could be given by the statement: **A** the student is male.

For some students this will be TRUE, and for them $A = 1$. For others it will be FALSE and for them $A = 0$, or equally it can be expressed $\bar{A} = 1$ (that is the *inverse* statement, the student is *not* male, is TRUE). I can group all the dots representing male students together, draw a closed curve round them, and label the space inside it **A** (meaning $A = 1$ inside); and the space outside it **Ā** (meaning $\bar{A} = 1$ outside) as in Figure 31. This is called a Venn diagram.

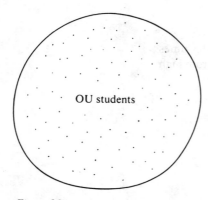

Figure 30

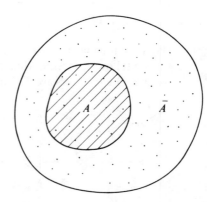

Figure 31

Another classification might be: **B** the student is employed as a diver. Again there will be some for whom this is true, $B = 1$, others for whom it is false, $B = 0$ or $\bar{B} = 1$; again draw a closed curve to separate the two subsets, as in Figure 32. In this case the two classifications would overlap, producing the combined diagram Figure 33. The diagram has now been divided into four regions, which have been labelled **AB**, **A$\bar{\text{B}}$**, **$\bar{\text{A}}$B** and **AB** (all four expressions of the form **$\bar{\text{A}}\bar{\text{B}}$** with and without inversion bars). Why can we do this? Look at the region labelled **AB**; this represents male students who are employed as divers; that is, it is a subset containing those students for whom **A** = 1 AND **B** = 1.

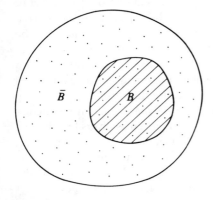

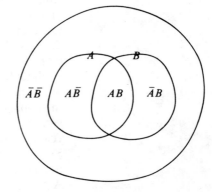

Figure 32

Figure 33

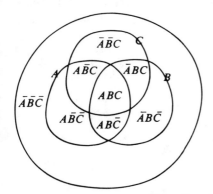

Figure 34

The notation for the AND function is **AB**; for this subset, with **A** = 1 and **B** = 1 **AB** will also = 1. On the other hand the group labelled **$\bar{\text{A}}\bar{\text{B}}$** represents female students who are not employed as divers. For them **A** = 0 and **B** = 0, which is the same thing as **$\bar{\text{A}}$** = 1 and **$\bar{\text{B}}$** = 1, so **$\bar{\text{A}}\bar{\text{B}}$** = 1. Similar arguments apply to the other two subsets **A$\bar{\text{B}}$** and **$\bar{\text{A}}$B**. If the subset labels **AB**, **A$\bar{\text{B}}$**, **$\bar{\text{A}}$B** and **$\bar{\text{A}}\bar{\text{B}}$** are expressed in terms of the values of **AB** which would make them respectively equal to 1, they will of course be 11, 10, 01 and 00, the usual four values of a two-bit code. Similarly, if there is a third classification: **C** the student is less than 35 years old, there will again be **C** = 1 and **C** = 1 subsets, forming a composite diagram as in Figure 34, using the same argument

53

about the AND function to explain the labelling of the various regions. Again, the maximum number of regions into which the diagram can be divided is eight, the same as the number of values of a 3-bit code.

So it goes on, the diagram being divided into smaller and smaller regions as the number of classifications, or the number of bits in the code, is increased. As the labels **ABC**, **ABC̄**, and so on, represent the minimum regions of the diagram which can be distinguished, they are called *minterms*; as you have seen they can be related to the various values of a binary code, or addresses in a truth table. In a practical problem, there may be some minterms which do not refer to anything because there are no members of the population in that category. Thus, it may be that in a particular year there are no female students employed as divers less than 35 years old (the subset **ĀBC̄**). Read that again if it seems obscure! Such composite classifications are sometimes difficult to disentangle in words; and this is for only three classifications. So the **ĀBC̄** minterm, or the corresponding code 001, describes a subset of students with no members. Thus, although an n-bit code can describe 2^n states, corresponding to 2^n minterms, not all of them need refer to anything, as you saw in the case of BCD codes for example.

A	B	0	AND A B	A B̄	A	Ā B	B	XOR ĀB + AB̄ A ⊕ B	OR A + B
0	0	0	0	0	0	0	0	0	0
0	1	0	0	0	0	1	1	1	1
1	0	0	0	1	1	0	0	1	1
1	1	0	1	0	1	0	1	0	1

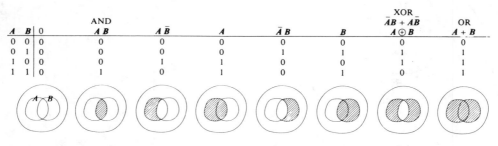

(each function in the bottom row is the inverse of the one above it)

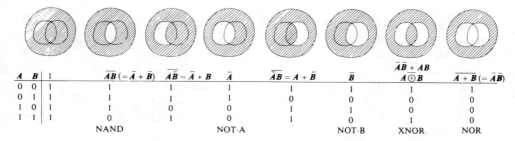

A	B	1	ĀB̄(= Ā + B̄)	AB̄ = Ā + B	Ā	ĀB = A + B̄	B̄	ĀB̄ + AB A ⊙ B	Ā + B̄(= ĀB̄)
0	0	1	1	1	1	1	1	1	1
0	1	1	1	1	1	0	0	0	0
1	0	1	1	0	0	1	1	0	0
1	1	1	0	1	0	1	0	1	0
			NAND	NOT-A		NOT-B		XNOR	NOR

Figure 35   The 16 logic functions of two variables

54

You may also find that the standard logic functions AND, OR, and so on, are easier to remember in Venn diagram than in truth table form. They are of course equivalent: you have seen how each 1 in the output column of a truth table corresponds in Boolean notation to one minterm (that is to one minimum region of the Venn diagram). So in general the output can be expressed as a sum of minterms, or as several regions of the Venn diagram. Figure 35 shows the 16 possible logic functions of two variables in Venn diagram form, the shaded region representing the combination of inputs giving the corresponding output.

CHAPTER 3

Sequential Logic

Introduction

So far we have looked at the idea of formulating simple problems in terms of a binary truth table. The truth table was shown to be a complete and systematic statement of the problem, providing the basis for a practical solution using digital combinational logic devices. However, the discussion was restricted to a particular type of problem. It was understood that a certain combination of binary digits would appear at the outputs of a device whenever a particular combination was applied to the inputs, IRRESPECTIVE OF THE PAST 'HISTORY' OF ITS INPUT. In other words, for combinational logic, the present output depends only on the present input and is not influenced by what has gone before.

In this chapter we want to introduce you to the idea of sequential logic circuits in which the inputs and outputs have the form of SEQUENCES of binary numbers. The behaviour of sequential logic circuits is not determined by a single input applied in isolation; THE PRESENT OUTPUT IS INFLUENCED BY THE HISTORY OF ITS PAST INPUTS AS WELL AS BY ITS PRESENT INPUT. The use of the word 'sequence' immediately implies some time-dependance in the circuit, 'the input is going through a sequence', in other words, the binary input is changing with time. Sequential logic circuits can be designed both to modify and to generate sequences of binary numbers. Application examples include counting of events and the timing of operations for the sequential control of industrial processes. The counting circuit provides a good introduction.

A basic sequential circuit: the binary counter

Electronic computers find applications in industry and in laboratories whenever the process or experiment involves the registration of individual *events*, for example, the counting of objects as they pass a particular point on a conveyor belt. When using electronic counters, the individual events must first be converted to a series of voltage pulses in order to activate the counter. In the conveyor belt case, this could be achieved by passing the objects between a photoelectric cell and a light source. Like its mechanical counterpart, the electronic counter can be reset by the operator to give an initial output of zero. At any later time the output of the counter indicates the number of events which have occurred since the count was first started.

In many applications the output of the counter is used to drive a digital display giving a denary reading of the total number of counts accumulated. For the moment we should study the case where the total count is provided in the form of a binary code.

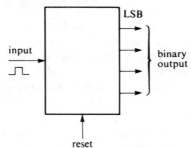

Figure 36   A binary counter

Figure 36 shows a block diagram of a simple *binary counter*. The *input* is provided in the form of a series of voltage pulses. The *output* appears in the form of a binary word which changes on the arrival of successive input pulses. The counter is initially reset to zero by the operator, this would normally be done by applying a voltage pulse of short duration to the *reset* input, to produce 0000 on the output lines. The output of the counter then changes on the receipt of successive input pulses. In a *natural binary counter*, the output word changes progressively in accordance with the natural binary code; that is, it follows the sequence

$$
\begin{array}{cccc}
0 & 0 & 0 & 0 \\
0 & 0 & 0 & 1 \\
0 & 0 & 1 & 0
\end{array}
$$

and so on up to the maximum count, 1111.

This is a 4-bit binary counter which generates a sequence of 16 binary numbers before reaching its limit. Like the mileometer in a motorcar, which resets to zero on reaching 99999, the electronic counter may also reset in the same way, depending on the application. The binary counter is one form of sequential logic circuit. First of all, the circuit generates a *sequence* of outputs in response to successive inputs. Secondly, at any stage in the sequence the *next* output word is uniquely determined by the *present* output word. Finally, the counter has the ability to 'remember' the effect of an input pulse. This ability sets it clearly apart from the combinational logic circuits discussed earlier.

Figure 37 shows a typical logic pulse; the convention for this chapter shows that a low level corresponds to logic '0' and a high to logic '1'.

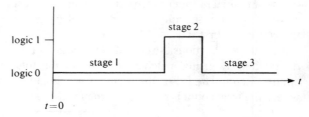

Figure 37   A logic pulse

Suppose a pulse of this nature is applied to the input of a combinational circuit. If we choose to ignore the propagation delay of the circuit, discussed in the last chapter then the output at any time will depend only on the input present at that time. In particular, the output during stage 3 in Figure 37 will be identical to the input during stage 1. In other words, stage 2 might never have occurred.

In contrast, the basic requirement in the binary counter is to register the fact that a pulse has occurred. This function is provided by *bistable* logical devices. A bistable device has two stable states, a *set state* defined when its output is a 1 and a *clear state* defined when its output is a 0. The output is usually in the form of a voltage which distinguishes between the two possible states of the device.

A bistable device can be made to change state when a pulse is applied to one of its inputs. This property gives rise to the commonly used term *flip-flop*, reflecting that the device can *flip* into one stable state or *flop* back again to the other on command. Therefore, the flip-flop can be regarded as a basic *memory device:* it can remember the effect of an input *after* the input has been removed.

Sequential logic circuits always incorporate devices with memory, but before any aspect of design can be discussed, the properties and characteristics of memory devices will be defined in the next section.

bistable
set state   clear state

flip-flop

memory device

58

Memory devices: flip-flops and registers

The *D*-type flip-flop

D-type flip-flop

The most simple flip-flop of practical importance in logic design is the *D-type flip-flop*. This example will serve to introduce several concepts which are applicable to memory devices in general. It is not necessary to understand the internal workings of the *D*-type flip-flop, but it is important to understand its terminal properties. Figure 38 shows the device as a black box.

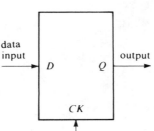

Figure 38   The D-type flip-flop

The *D*-type flip-flop has a *data* input *D*, an output *Q*, and the pulses are applied to the *clock* input *CK*. The reason for using the word 'clock' in this context will become clearer later. In combinational logic circuits the truth table gives a complete summary of the relationship between all possible inputs to a circuit and the resulting outputs.

> Is it possible to draw up a truth table for a flip-flop?

> In general, yes. Flip-flops are relatively simple devices with a limited number of input-output possibilities. However, the truth table must take into account the effect of applying a pulse on the clock input. To avoid confusion the truth table for a flip-flop is usually called a characteristic table.

The characteristic table for the *D*-type flip-flop is very simple. It is shown in the table below. The output following a clock pulse depends only on the input *D* at the time the clock pulse is applied.

Input *D*	Output *Q* after receipt of a clock pulse
0	0
1	1

memory element

In circuit design the *D*-type flip-flop is often used as a basic *memory element* for the short-term storage of a binary digit applied to its data input. The binary input present at the time the clock pulse is applied is transferred to the output and held there until another clock pulse is applied.

set-direct
clear-direct

A practical flip-flop, as shown in a manufacturer's catalogue, usually has additional connections including an extra output, $\bar{Q}$, the logic complement of Q. A more typical block diagram is shown in Figure 39. The inputs labelled S_D and C_D are the *set-direct* and *clear-direct* inputs. A pulse applied to these inputs will switch the output Q to its *set* state ($Q = 1$) or to its *clear* state ($Q = 0$), respectively. They are commonly used to define an *initial state* before the arrival of the first clock pulse.

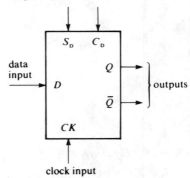

Figure 39   A practical D-type flip-flop

The clocked register

memory register

Two or more basic memory devices can be grouped together to form a *memory register* or, more usually, a *register*. An *n*-bit register is capable of storing an *n*-bit binary word. A register which is activated from a clock pulse applied simultaneously to all the individual elements is known as *clocked register*. A typical clocked register is shown in Figure 40.

clocked register

In Figure 40 the input word is applied on a set of wires, one wire for each bit of the input word. Although the input word may be changing with time (for example, it might be supplied from the output of a binary counter) the register makes no attempt to

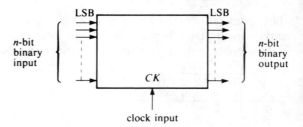

Figure 40   An n-bit clocked register

60

'memorize' any of the sequentially applied inputs until activated by the clock input. Let us look as this 'activation' in more detail.

The transfer to memory is initiated by the change, or *transition*, of the clock input between logic levels; some registers and clocked memory devices are designed to respond to a 1 to 0 transition; others to a 0 to 1 transition. The ability to 'latch' on to a binary number in a controlled way has given rise to the alternative name *latch*, for the clocked register. I shall not use this term to describe the device, although I shall find it convenient to refer to the 'instant of latching'. It is worth noting that there are basically two types of clock pulse which might occur in digital circuits as shown below.

latch

Figure 41 defines the *leading edge* and *trailing edge* of the clock pulse. A device which latches on a 0 to 1 transition will be latched on the leading edge of pulse type A and on the trailing edge of pulse type B. Hence a memory device will only operate satisfactorily when the input data are stable at the instant of latching.

In view of this, the timing sequence for a particular device is always closely specified by the manufacturer. Figure 42 shows a typical example. This particular device is actuated by a 0 to 1 transition of the clock input. The clock waveform is shown with a finite *rise time*, reflecting that the voltages in a practical circuit cannot change instantaneously.

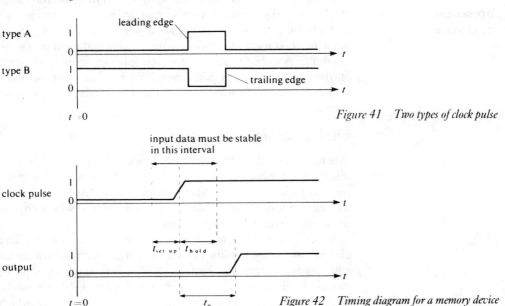

Figure 41   Two types of clock pulse

Figure 42   Timing diagram for a memory device

The data input must be present for a certain minimum time before the transition is applied. This time is referred to as the

set-up time

hold time

set-up time ($t_{\text{set-up}}$). The input data must also remain fixed for a minimum time after the transition has occurred. This time is referred to as the *hold time* (t_{hold}). The *propagation time* to the time that the output changes state is also shown in Figure 42. Another device might be specified in a similar way, only, this time, referred to a 1 to 0 transition of the clock. In either case they are referred to as *edge-triggered* devices.

When a circuit designer has control over the generation of the clock pulses and a choice of clocked registers he may have to decide whether to arrange for latching on the leading edge or on the trailing edge of the clock pulse. When the clocked register forms part of a sequential logic circuit, the timing diagram provides a summary of the timing requirements which must be satisfied by the *overall circuit* to ensure correct operation.

The 'state' of the register

When referring to the individual memory elements of a register as being in one of two possible states it is equally meaningful to enquire as to the *state of the register*. This would be given by identifying the state of the individual elements, say with a voltmeter, and could be expressed as an *n*-bit binary word, where each bit indicates the 'content' of each element.

present state
next state

Because clocked registers are always used in association with clock pulses the terminology has expanded to include *present state* and *next state,* that is, the next state to appear after the next clock pulse has taken effect. Of course the present state and next state are not *necessarily* different. In the simple binary counter, the state of the output changes with every input pulse while in other sequential circuits the output sequence may contain several *repeated* states.

Binary voltage waveforms

We have seen that the output of a clocked register is an *n*-bit binary word which changes on the receipt of successive clock pulses. To imagine these 'successive words' or 'successive states' as being nothing more that 'blocks of numbers' which change from time to time, however, is dangerous.

When a memory device changes state there is a corresponding change in *voltage* on its output line; logic circuits work with voltages and currents and interact with other circuits by virtue of *changes* in their voltages and currents. Figure 43 shows how a sequence of states can give rise to *binary voltage waveforms*, which can be measured in the circuit and displayed on an oscilloscope. Figure 43 assumes that the sequence was 'clocked' at regular intervals.

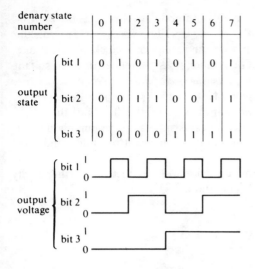

denary state number		0	1	2	3	4	5	6	7
output state	bit 1	0	1	0	1	0	1	0	1
	bit 2	0	0	1	1	0	0	1	1
	bit 3	0	0	0	0	1	1	1	1

Figure 43   Generation of binary voltage waveforms

You will see that each individual state is identified in the sequence with a denary state number. In the present case, where the sequence follows a natural binary code, the denary state number is equivalent to the weighted binary output, but this is not always the case. The diagram shows the binary waveforms generated by this succession of states; one waveform for each individual bit of the output word.

More about flip-flops: the *J–K* flip-flop

This section would not be complete without discussing the properties of another type of flip-flop, the *J–K flip-flop*.

J–K flip-flop

The versatility of the *J–K* device results from the fact that it has *two* data inputs as shown in Figure 44. The other inputs, S_D and C_D, have the same functions, as defined for the *D*-type flip-flop.

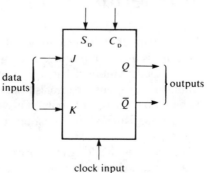

Figure 44   The J-K flip-flop

To show the characteristic table for the *J–K* device in the same

63

Table 2 Characteristic table for a J–K flip-flop

Inputs $J\,K$	Output Q_{n+1}
0 0	Q_n
0 1	0
1 0	1
1 1	$(\overline{Q_n})$

form as it appears on manufacturers' data sheets, Table 2 introduces two new symbols Q_n and Q_{n+1}. The first, Q_n, represents the output of the flip-flop after the receipt of n clock pulses. It may therefore be 0 or 1. The symbol, Q_{n+1}, is defined as the resultant output Q of the flip-flop *after the receipt of the next clock pulse.*

Following earlier comment on the 'state' of a device, if Q_n represents the *present* output state of the flip-flop, then Q_{n+1} represents its *next* state. These symbols are commonly used in manufacturers' literature so it is important to understand their meaning.

Table 3 is an expanded version with comments to further clarify the use of the Q_n, Q_{n+1} notation.

Table 3 Characteristic table of a J–K flip-flop

Inputs $J\,K$	Resultant output after clock pulse, Q_{n+1}	Meaning
0 0	Q_n	No change in the output Q
0 1	0	Output Q *always* 0, irrespective of previous state of output
1 0	1	Output Q *always* 1, irrespective of previous state of output
1 1	$(\overline{Q_n})$	Output Q *always changes state irrespective of previous state*

General properties of sequential circuits: counters and sequencers

The remainder of this chapter will discuss applications in which a designer is asked to produce circuits for counting and sequence generation, circuits which generate or modify a sequence of binary numbers in response to signals from the outside world. To do this he has two basic sets of 'tools' at his disposal; combinational logic circuits (or ROMs) and memory devices such as registers or flip-flops.

To set about the design task in a systematic way it is necessary to understand something about the overall form of a sequential logic circuit. The usual textbook approach is to do this in terms of the 'general' sequential circuit model which was devised as a theoretical basis for systematic design. The problem is that this approach is abstract and tends to be confusing. The model has a number of inputs and outputs and allows a virtually complete interconnection of the combinational and memory elements. In order to analyse a familiar device in sequential logic terms, we have chosen the J–K flip-flop.

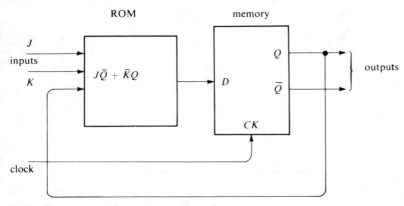

Figure 45   The J-K flip-flop shown as a sequential logic circuit

The *J–K* flip-flop as a sequential logic circuit

It should be clear to you by now that the J–K flip-flop functions as a self-contained sequential circuit. A competent designer could undertake the task of devising a sequential circuit with the same characteristic table as the J–K flip-flop, using an interconnection of combinational and memory devices. His design solution could look something like Figure 45. The circuit requires a combinational block specified by a Boolean expression, shown as a ROM, and a D-type flip-flop used as a basic clocked memory element. The circuit embodies an important feature, that is, *feedback* from the output of the memory to the input of the combinational block. The ROM has three inputs. One of these is *internal* from the memory device, the other two are *external*, supplied by the J and K data inputs. At any time the output from the ROM depends on the inputs *at that time*. The *present* ROM output therefore depends jointly on the present external inputs *and* the *present* state of the memory device.

A truth table for the ROM is shown in Table 4.

On the receipt of a clock pulse the *present* D input will be transferred to the output of the flip-flop. Using this fact, together with the truth table for the ROM, data can be drawn up as in Table 5 on p. 66.

This serves to emphasize a fundamental property of sequential logic circuits in general, namely that the *next* output state is

Table 4 Truth table for the *J–K* ROM

Inputs $J\,K\,Q$	$D = J\bar{Q} + \bar{K}Q$
0 0 0	0
0 0 1	1
0 1 0	0
0 1 1	0
1 0 0	1
1 0 1	1
1 1 0	1
1 1 1	0

uniquely determined by the *present* external inputs and the *present* state of the memory devices.

Table 5 Present state-next state description of the *J–K* flip-flop

Present state		Next state
Inputs *J-K*	Memory *Q*	Output
0 0	0	0
0 0	1	1
0 1	0	0
0 1	1	0
1 0	0	1
1 0	1	1
1 1	0	1
1 1	1	0

The designer of the *J–K* type circuit presumably knew about this basic property in advance. He would also have been aware that most sequential logic circuits incorporate feedback between the outputs and inputs of the memory block and the combinational block. These factors give the key to systematic design. The first step would be to take the characteristic table for the *J–K* flip-flop and use the 'present state-next state' terminology to express it in the form of the table above. He would then arrange for the *present-state* information to provide an address to the ROM. The ROM would be programmed to supply the *next* output state to the input of the memory device in readiness for the next clock pulse to **next-state ROM** arrive. In this type of application the ROM is called the *next-state ROM*. Once the truth table for the next-state ROM has been specified the design is virtually complete.

Many important sequential circuits can be designed in a more or less routine fashion provided that the correct initial steps are taken when the required sequence of states is analysed. In general, most sequential logic circuits incorporate a memory register rather than a single memory device. The memory register may not always supply the output of the circuit directly – in some cases, the memory output may be converted to a different output code using **output ROM** an *output ROM* – in others, only a part of the memory register may be used to supply the output of the circuit. In general, therefore, the present external output will be different to the present state of **internal state** the memory devices. This represents the *internal state* of the sequential logic circuit, reflecting the fact that it may be impossible to identify the state of the *entire* memory when the circuit is presented as a (physical) black box. Thus in a sequential logic circuit the *next* output state is uniquely determined by the present external inputs and the *present internal* state of the overall circuit.

Counting circuits

A *counter* is a sequential logic circuit which generates a prescribed series of output states upon the application of pulses to its clock input. In the light of comments in earlier sections you will appreciate that the basic sequence is *finite* in length, but that it may repeat itself (remember the analogy of the mileometer resetting at 'full count'). You have seen that a *binary counter* produces a sequence of states which follows the natural binary sequence. If a 3-bit binary counter is initially reset to an initial state of 000 and is allowed to repeat it will generate the sequence shown in Figure 46 on the receipt of successive clock pulses. The sequence follows the arrows from its *present state* to its *next state* only on the application of a *count pulse* to its clock input. A counter which generates a sequence of *n* states before repeating itself is known as a *modulo-n counter*. The counter of Figure 46 is a modulo-8 counter. Counters can be designed to operate with almost any sequence and any modulo. For example, the *decade counter* or *modulo-10 counter* follows a sequence of 10 states before returning to its initial state while a *modulo-16 counter* operates with 16 possible states. Cyclic codes such as the Gray code can be produced in sequential form using the appropriate counter configuration.

Counter design: basic principles

The operation of a counter is fully defined in terms of the output sequence generated by the repeated application of clock pulses. The design problem is to specify the combinational logic and clocked register and their interconnections in order to achieve a given sequence. The required sequence of states is usually presented in the form of a *state table*. The table below shows the state table for a 2-bit binary counter. A denary state number is used to identify the individual output states in the basic sequence which is allowed to repeat as often as required.

Table 6 State table for 2-bit binary counter

Denary state number	Output state
0	00
1	01
2	10
3	11

Sequential logic problems are *always* clarified by using the 'present state—next state' terminology. By adding an extra column,

modulo-*n* counter

**decade counter
modulo-10 counter**

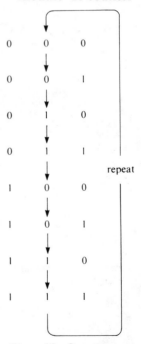

Figure 46   State sequence for a 3-bit natural binary counter

67

the table shown below holds the key to the practical implementation of the counter shown in Figure 47.

The implementation incorporates the principle of feedback from the memory devices in the clocked register to the input of the ROM. In turn the ROM supplies the next state information to the clocked register. The ROM is therefore being used as a next-state ROM with a truth table as shown in Table 8.

Table 7 'Present state-next state' table for a 2-bit binary counter

Denary state number	Present output state	Next output state
0	0 0	0 1
1	0 1	1 0
2	1 0	1 1
3	1 1	0 0

Table 8 Truth table for the next-state ROM

Inputs	Outputs
0 0	0 1
0 1	1 0
1 0	1 1
1 1	0 0

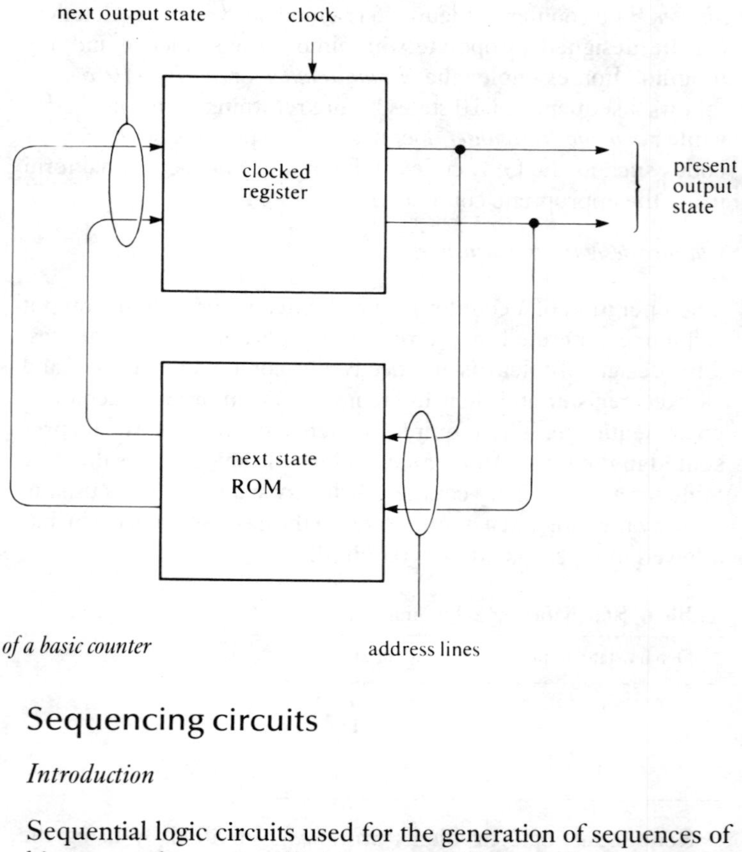

Figure 47   Implementation of a basic counter

Sequencing circuits

Introduction

sequencers

Sequential logic circuits used for the generation of sequences of binary numbers are called *sequencers*. Their applications in the

68

general field of digital control systems are many and varied. Briefly, any process which consists of a well defined sequence of events (for example, an automatic production line, or the operation of automatic machine tools) can, in principle, be taken from step to step under the control of a digital or logical sequencer. Because sequencers and counters appear to be very similar in their function it is important to identify some of the differences between them.

In a counter, the 'input' is in the form of pulses (count pulses) applied to the clock line. These *may* come from a master oscillator or clock, but in general, the sequence of states in a counter is advanced by the application of *external* pulses to the clock line. These external pulses may arrive in more or less random fashion, for example, when counting items on a production line. By way of contrast, when controlling a sequence, the time spent in each state is usually a major design consideration and this is almost invariably determined by a system clock with a well defined period.

There is a second consideration involving the *content* of the sequences generated by the two types of circuit. A counter is normally used in such a way that the present state of its output can tell the user how many count pulses have occurred since the counter was first 'started'. This implies that the counter clocks through a sequence of *individually distinguishable* output states (note: this is *not* a definition of a counter).

When controlling a process with a sequencer, however, it may be necessary to hold the outputs at the same value for several clock periods or for the same output state to appear in different parts of the sequence. By assigning a state number to each stage of the sequence we can keep track of what is happening. One way of doing this is to include a counter within the sequencer to record the arrival of each clock pulse. At any time, therefore, the counter indicates the number assigned to the present state of the sequence. The counter output can be converted to any required output state using combinational logic.

A typical arrangement is shown in Figure 48 (p. 70).

This circuit can be the starting point in the design of almost any sequencer. It provides an example of a circuit in which the present *output* state may be quite different from the present *internal* state held on the counter memory circuits. In this arrangement the binary counter is said to function as an *internal state register*.

internal state register

More complicated sequencers

In process control applications it may be necessary to change the generated sequence in response to signals fed back from the operation. The required change could be to extend the duration of a particular state or perhaps to divert from the main sequence to a secondary or auxiliary sequence for a prescribed time.

Sequencers which respond in this way are known as *conditional
sequencers*, and some design considerations will be met later on in
the chapter.

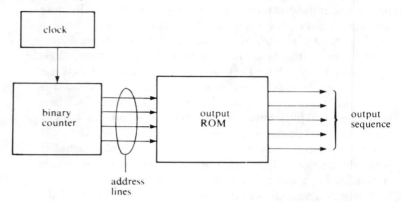

*Figure 48   Implementation of a simple sequencer using a binary counter as an
internal state register*

Synchronization in sequential logic circuits

A clocked register can only operate satisfactorily when the binary
data inputs and the clock pulses satisfy certain timing constraints.
When several clocked registers and combinational circuits, each
with its own propagation delay, are interconnected to form a
sequential logic circuit it is clear that the question of timing
assumes major importance if the circuit is to perform to specification.

In circuits of moderate complexity it is commonly the case that
the data input to one clocked register or flip-flop is supplied by the
output of another clocked register or flip-flop. Figure 49 shows
part of a sequential circuit involving two D-type flip-flops. If the
state of flip-flop 1 is to be transferred to flip-flop 2 it is essential
that both flip-flops have completed any previous state change
before the transfer is made. For example, if flip-flop 1 is 'settling'
into a new state when a clock pulse is applied to flip-flop 2 the
circuit will not operate in a predictable fashion. This may leave the
output Q_2 in an incorrect condition, which will divert the overall
sequential circuit from its design sequence. One way in which
timing problems can be minimized is to ensure that all the memory
devices in the circuit change state *simultaneously*, and we will be
primarily concerned with sequental circuits of this type. They are
**synchronous
sequential circuits** known as *synchronous sequential circuits*, and have a single clock
input which is common to all the memory devices. In a synchronous
circuit the flow of data can only take place following the receipt of

70

a clock pulse. The designer responsible for the circuit must ensure that the clock period is sufficiently long compared with the 'settling' time of the overall circuit, which will not be less than the propagation time of the slowest component device. Given this, the circuit conditions are quite *static* before the next clock pulse arrives and the data inputs to all the memory devices are in a well defined state.

The idea of synchronous operation is an important one; look at the timing sequence for an arrangement like Figure 49 in some detail. Assume that the data input to flip-flop 1 is temporarily 'fixed' at logic level 0 as shown in Figure 50. The flip-flops have initial states of 1 and 0 and the clock input is applied to both flip-flops simultaneously.

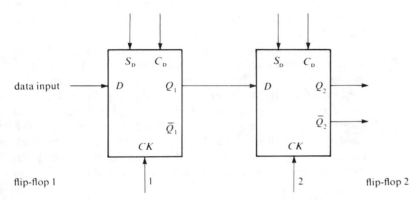

Figure 49   Part of a sequential logic circuit

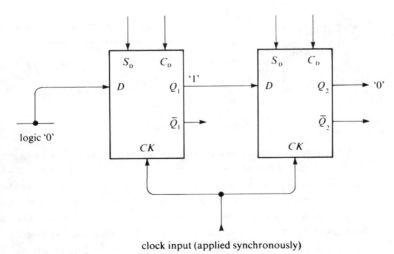

clock input (applied synchronously)

Figure 50   Synchronous operation of two flip-flops, showing the initial conditions

71

Figure 51 shows the timing sequence when the flip-flops latch on the 0 to 1 transition of the clock pulse. You will see that we have taken account of the propagation delay, t_p, of the flip-flops in Figure 51. The input to flip-flop 2 is stable at the instant of latching, so there will be no problem in satisfying the set-up time requirements of flip-flop 2. You can refer back to Figure 41 which shows the timing diagram for the flip-flops. But what about the hold time of flip-flop 2? If you examine the timing diagrams carefully you will see that correct operation can be guaranteed only when the propagation delay of flip-flop 1 is greater than the hold time of flip-flop 2. If you repeat the timing diagram with the flip-flops actuated on the 1 to 0 transition of the clock you will find that the same considerations apply.

When the two flip-flops are of identical type this will not normally be a problem, but in general system design it may be desirable to 'mix' flip-flops of different speeds for reasons of economy. In such a case, the timing characteristics of each individual device must be given careful consideration. This particular timing problem could be overcome by using flip-flops in which the hold time was effectively zero. Another flip-flop timing sequence where this is indeed the case is the *master-slave timing sequence* shown in Figure 52. Figure 52 shows that devices operating on the master-slave timing sequence respond on both edges of the clock pulse. The input enters the flip-flop when the clock goes high, then transfers to the output when the clock goes low.

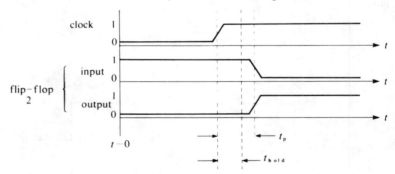

Figure 51   Timing diagram for Figure 50

An important requirement is that the input data should be stable for the duration of the clock pulse. The set-up time is therefore equal to the width of the clock pulse; the minimum width for proper operation is always given on the data sheet for the device. The data can be removed immediately after the latching transition, so the effective hold time is zero for this type of device. The propagation delay is measured from the latching transition as shown in Figure 52.

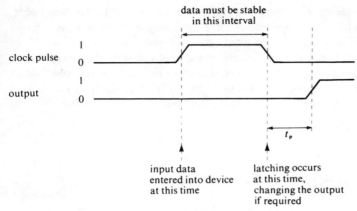

Figure 52   Timing diagram for the master-slave timing sequence

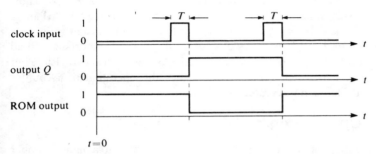

Figure 53   Timing diagram for the sequential circuit of Figure 45 containing a flip-flop and a ROM

To summarize, therefore, a synchronous sequential circuit can be made to operate successfully when the memory devices are edge triggered, *provided* that proper attention is given to the timing requirements of each individual device. Many timing problems can be overcome by using master-slave devices in which the effective hold time is zero. In all cases, however, the propagation time of any combinational devices in the system must also be taken into account.

Example: Suppose the sequential circuit in Figure 45 has the $J-K$ inputs connected to logic level 1. The D-type flip-flop operates on the master-slave principle. Figure 53 shows a timing diagram for the circuit when the flip-flop is initially cleared to $Q = 0$. Note that the time-scale is too long to show the effect of propagation delays in the circuit.

Finally, consider the design of synchronous sequential circuits, namely: how can *externally* applied inputs change in synchronization with a system clock? This may not always be possible: in the worst case the problem of *interfacing* two independent synchronous systems which are operating at different clock rates may occur. There are nevertheless a great many sequential logic circuits which operate with fixed external inputs, where the sequence of states is advanced purely by the application of clock pulses. You should appreciate by now that many digital counters are included in this category. In other types of circuit, such as conditional sequencers, the external inputs may be applied infrequently or irregularly.

Design of counters

Introduction

We have discussed the basic concept of a counter and showed how a very simple counter could be implemented using a ROM and a clocked register. Now we can extend the discussion to include several different types of counter of practical importance. In all cases the design is based on a ROM and clocked register, but there is a section which describes the operation of a counter available in integrated circuit form.

Up-counters and down-counters

up-counter
down-counter

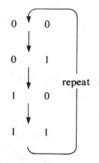

Figure 54   State sequence for a 2-bit counter

The idea of counting 'up' from a low number to a higher number and, conversely, 'down', has carried over into digital counters, so that it is now common practice to indicate whether an *up-counter* or a *down-counter* is required for a particular application. In general an up-counter generates a sequence of states where the equivalent binary number *increases* from count to count; the down-counter *decreases* its binary output from count to count. Using the simple state diagram introduced earlier, this time, for a 2-bit natural binary sequence, the direction of the arrows in Figure 54 tells us that the counter is counting 'up' from its initial state of 00. When the counter reaches 11 it must make an apparent 'down' transition to 00 in order to repeat: The fact is, of course, that the counter is continuing to count 'up', but there are not enough output bits to show it.

Setting up the initial state of a counter

preset number

It is common practice to arrange for a counter to reset automatically to its desired initial state following switch-on. In general, of course, the desired initial state might be a number other than 00 . . . 0. In establishing the required initial state the counter is said to be *preset* with *a preset number*.

A *preset counter* is a counter with the facility to accept a wide range of preset numbers which can be selected and *loaded* by the user. Clearly these remarks apply equally to up-counters and down-counters.

Up-down counters; reversible counters

When introducing the topic of sequencers we have seen that, in some cases, the output sequence might be modified in response to additional inputs supplied to the circuit. This is classified as a *conditional sequencer.* The *up-down counter* or *reversible counter* is a type of conditional sequential logic circuit; it is a counter which counts either 'up' or 'down' according to the state of a *direction control input.*

up-down counter
reversible counter

The design principles of an up-down counter can be illustrated quite adequately with a 2-bit binary counter, the extension to counters of greater capacity is really only a matter of scale. There are two possible sequences shown in Tables 9 and 10 using 'present state–next state' terminology. One possible solution is to make use of a ROM with three address lines as in Figure 55. One of the address lines forms an external input to the combinational logic. It is set to 1 for counting UP and 0 for counting DOWN.

Table 9 Counting UP

'Direction' input	Present output state	Next output state
1	0 0	0 1
1	0 1	1 0
1	1 0	1 1
1	1 1	0 0

Table 10 Counting DOWN

'Direction' input	Present output state	Next output state
0	1 1	1 0
0	1 0	0 1
0	0 1	0 0
0	0 0	1 1

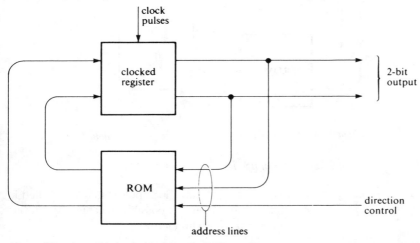

Figure 55   A possible implementation of a 2-bit up-down counter

You will see that the extra storage element is to be *leading-edge* triggered. If the other memory elements are *trailing-edge* triggered, this will ensure that the direction control takes effect *before* the input pulse is registered by the counter. In designing the circuit, account must be taken of the timing requirements of all the memory elements and also the propagation time of the ROM.

The direction control input will be latched successfully if it satisfies the set-up time of the extra storage element on the 'next' clock pulse. The new 'direction bit' input to the ROM appears after the propagation time of the storage element has elapsed. The correct 'next state' code appears on the input to the clocked register after a further delay given by the ROM propagation time. The state of the circuit is then fixed until the clock pulse makes the transition from 1 to 0. The next-state information is then transferred to the output of the clocked register, the indicated count being either 'one-up' or 'one-down' on the previous count, depending on the direction control input.

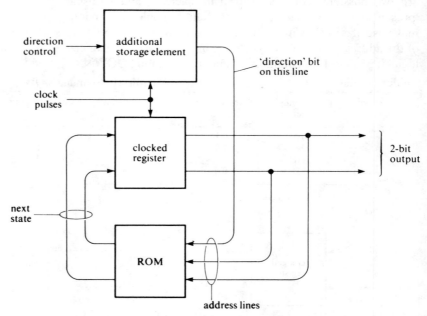

Figure 56 Improved implementation of the up-down counter

Synchronization in up-down counters

In a typical counter the clock pulses may be applied in random fashion. How is it possible to achieve synchronization with the direction input?

The solution is to apply the direction input in such a way

that it is clocked into the register by the next clock pulse. This approach requires expansion of the register by one bit, by the use of an extra binary storage element. The arrangement is shown in Figure 56, together with a timing diagram, Figure 57.

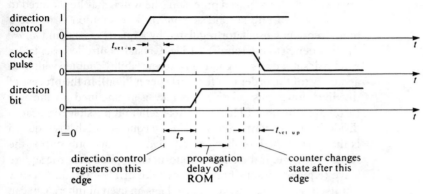

Figure 57   Timing diagram for Figure 56

Divide-by-*n* counters

In introducing the topic of *divide-by-n counters* we want to shift the emphasis to looking at the binary waveforms generated by the counter. This is because the main application of divide-by-*n* circuits in modern electronics is for *frequency division* in which the repetition frequency of a binary waveform is reduced from, say, *f* to *f/n*. This can be illustrated by Figure 58, which shows the binary waveforms generated by a modulo-4 or 2-bit binary counter. The counter has an initial state 00 and resets after the fourth clock edge. Comparing the clock rate and the repetition rates of the two output waveforms you will see that the circuit gives the choice of frequency division by two or by four depending upon which bit is used as the output.

divide-by-*n* counters

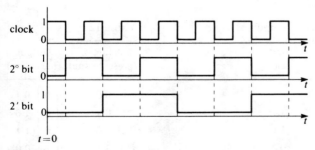

Figure 58   Divide-by-2 and divide-by-4 operation with a binary counter

A familiar example which makes use of a natural binary counter is found in the digital watch. The problem here is to provide a compact, precision oscillator of high stability to produce a sequence of pulses at one second intervals: Low-frequency oscillators are not compact (they involve the use of large capacitors) nor do they operate with the required precision (the watch may be required to be accurate to one second a month). The solution is to use a high-frequency oscillator regulated by a quartz crystal and *divide* the frequency to the required 1 Hz. Crystal-controlled oscillators can be designed to give good long-term stability coupled with high accuracy (of the order of a few parts per million). In the process of dividing down this relative accuracy is maintained all the way through to the final 1 Hz output. It is generally possible to extract a divide-by-n waveform from a modulo-n counter. All that is required is that one of the bits changes state *once* and once only during the basic sequence. If the clock waveform is supplied as a true square wave (equal durations high and low) the divide-by-n output may not itself be a square wave – even if n is an even number. You can see this by writing out the state sequence for the single decade BCD counter. In this case, the output from the MSB is 'low' for eight states and 'high' for the final two. If a square wave is required, therefore, it may be necessary to modify the next-state ROM. As an alternative, the counter could be used as a *state register* and supply the address to an *output ROM*, thereby turning the circuit into a simple sequencer.

Another example in this category is the provision of an output to indicate that the counter has reached the final state in its basic sequence. An output of this sort which appears for the duration of the final state is known as a *carry output*. The carry output is usually treated as an additional output over and above the normal n-bit output. The state sequence for a 2-bit binary counter with a carry output is shown in Table 11. The next-state ROM truth table is shown in Table 12. You will see that the MSB which contains the carry information has been blocked off.

The implementation is shown in Figure 59.

carry output

Table 11 2-bit binary sequence with carry output

Present state	Carry output	Next state
0 0	0	0 1
0 1	0	1 0
1 0	0	1 1
1 1	1	0 0

Table 12

Address (present state)	Output (next state)	
0 0	0	0 1
0 1	0	1 0
1 0	0	1 1
1 1	1	0 0

Counting over several decades

Modulo-10 or decade counters are the preferred type in cases where the count information is presented in the form of a numeric display or print-out. So far, however, the discussion has centred on the operation of a *single decade* counter which resets to its initial state after a count of 9. Let us look at the more general problem of counting over several decades so that several thousand or even millions of counts can be registered before the counter resets.

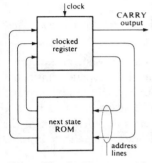

Figure 59   Basic counter with carry output

One possible approach would be to design a modulo-100 or modulo-1000 counter, that is a modulo-10^n counter to cover n decades. The practical solution is to *cascade* several decade counters, one for each decade as shown in Figure 60.

cascade

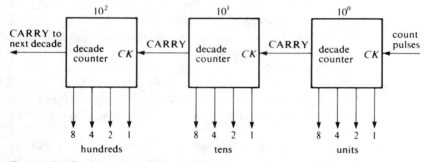

Figure 60   Cascade connection of several decade counters

If the counter is to be designed using a next-state ROM and register, how many lines would the ROM truth table contain?

The absolute *minimum* would be 10^n, say one millionn lines for a 6-decade counter. So the next-state ROM solution is impracticable.

The input to the complete assembly is applied to the 10^0 or units decade. The clock input to each of the subsequent decades is taken from the *carry* output of the preceeding decade. In this arrangement each decade counter often operates in a BCD code and is associated with one of the digits in a numeric display. The digits may be of the 7-segment type and may be connected to the corresponding BCD output via a 7-segment decoder. The least significant digit clocks up on every count while any other digit must wait until the preceding stage has clocked through 10 states before changing. The individual BCD counters provide a 4-wire output. An additional *carry* output is required on every tenth count pulse. When designing with ROM and clocked register therefore the ROM output consists of 5 bits.

Is the overall circuit shown in Figure 60 strictly a *synchronous counter?*

No; the individual decade counters may operate synchronously, but, except for the units decade, state transistions are not caused by the count pulse input but by the carry pulse from the previous stage. The effect of a carry pulse from the units decade therefore 'ripples' through the counter as the individual decades change state in rapid succession.

Ripple carry can give problems at high count rates, especially when the number of decades is high. The critical design factor is the propagation delay time of the individual decade counters, which in a TTL design may be of the order of 30 ns. In a 6-decade counter, therefore, the total delay between a count pulse at the input and the resulting change in state on the last decade (say when the total count changes from 199 999 to 200 000) is about 180 ns. For reliable operations at high speeds a fully synchronous connection of the cascaded counters is often preferred. The individual counters then change state simultaneously, and the total delay is reduced to that of a single counter. The technique is to apply the input count pulses to the clock inputs of *all* the decade counting stages. Additional gating is then arranged to enforce the following condition:

A counting stage will change state on the next clock pulse only when *all* the previous stages are displaying a 9. Many decade counters give a carry output which persists as a logic high level for the entire time that the counter displays a 9. In this case, the gating around each individual counter takes the form shown in Figure 61.

Clock pulses will take effect on a counting stage only when the previous stage and the least significant stage are displaying a 9. In turn, the previous stage can give a carry out only when *its* preceding stage is displaying a 9 and so on down the chain of cascaded counters. When BCD counters are purchased in integrated-circuit form the gating is often included within the counter package and the control inputs, P ENABLE and T ENABLE, are made available for cascaded operation. Figure 62 shows a cascade connection of four decade counters using the P ENABLE and T ENABLE inputs.

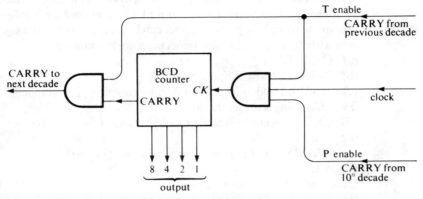

Figure 61   *Additional gating required on a BCD counter for synchronous cascaded operation*

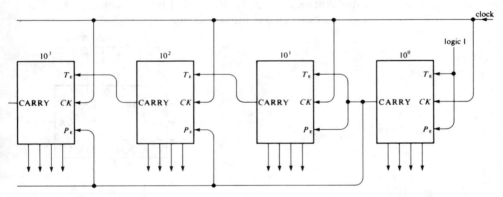

Figure 62   *Cascade connection of BCD counters using the T ENABLE and P ENABLE inputs to each stage*

Programmable counters

Introduction

If you look through the 'logic devices' catalogues produced by the major semiconductor manufacturers you will find an enormous variety of counters available 'off the shelf' in the form of integrated-circuit packages. There is not enough space here to review the field, taking into account the different types of counter and the various logic families (available and forthcoming), so a particular

81

programmable counter   device, an example of a *programmable counter*, is discussed. Figure 63 shows the essential pin configurations on the integrated circuit. This counter has the facility to enter a preset number to determine its initial state. The description 'programmable' refers to the fact that the counter can be connected to operate as an up-counter with any modulo up to a maximum of 16. The counter is fully synchronous and the preset number can be loaded in synchronization with the clock. The pin connections are as follows::

(a) OUTPUTS: Q_A, Q_B, Q_C, Q_D.

(b) PRESET INPUTS: P_1, P_2, P_4, P_8.

(c) LOAD: Held at logic 1 for normal counting. Taken to logic 0 when loading a preset number.

(d) CLEAR: A clear-direct input. Held at logic 1 for normal counting

(e) CARRY OUT: Goes to logic 1 when $Q_A = Q_B = Q_C = Q_D = 1$.

Gives logic 0 on all other counts.

(f) CLOCK: The counter generates its output sequence following the receipt of clock pulses in the usual way.

(g) TENABLE, P ENABLE: Counting is inhibited when either input is at logic 0. The CARRY OUT is gated with the T ENABLE input as described in the last section.

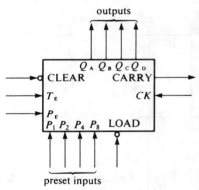

Figure 63   Pin connections for a programmable counter

Loading the preset count

The LOAD input is taken to logic 0 and the desired initial state is entered by connecting the appropriate PRESET input either to logic 0 or logic 1 as required.

Counting sequence

The counter output agrees with the PRESET input *after the next clock pulse* regardless of the state of the LOAD input. When the LOAD input is taken to logic 1, the count sequence begins at the preset state and counts up to 1111 following the natural binary code. The 1111 state is 'recognized' internally and the CARRY output goes to logic 1. The CARRY output can be taken through an inverter and fed back to the LOAD input enabling the preset values to be entered again. The counter then counts repeatedly from the preset number up to 1111, resetting on every CARRY pulse.

Figure 64 shows how this can be arranged in practice, using an inverter to reload the preset number. Figure 65 summarizes the overall timing requirements for the counter. The loading of the preset number is accomplished within a single clock cycle. Whenever the LOAD input is taken to logic 0, the preset number is entered synchronously with the next clock pulse. The counter output now agrees with the preset number and the CARRY output switches to logic 0, causing the LOAD input to go 'high' and enable the counting to proceed on successive clock pulses.

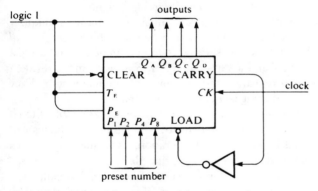

Figure 64   Using an inverter to load the preset number

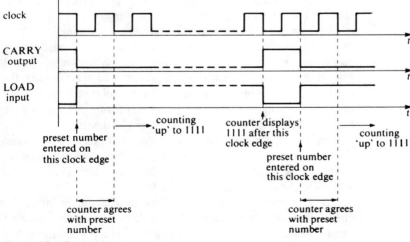

Figure 65   Timing diagram

83

Summary

We have described some of the properties and applications of *sequential logic circuits*.

Sequential logic circuits have the following properties:
(a) Their outputs depend on the history of their previous inputs.
(b) They incorporate logic devices with *memory*.

One type of memory element is the D-type flip-flop which is described by its *characteristic table*. Several D-type flip-flops can be combined to form a *clocked register* used for the short-term storage of a binary word. The *timing diagram* for a register or flip-flop summarizes the conditions which must be met in order to ensure correct *latching* of the input data. The *sequence of states* at the output of a register can be usefully discussed in terms of *binary voltage waveforms*. The *J–K flip-flop* has two data inputs. It is used extensively as a memory element in logic circuit design.

The *next* output state of a sequential logic circuit is uniquely determined by the *present* external inputs and the *present internal state* of its memory devices. Basic counters can be devised using a clocked register and a *next-state ROM*. *Sequencers* may provide a sequence with repeated states or states of unequal length. They can be designed using a counter as an *internal state register*.

In a *synchronous sequential circuit* clock pulses are applied to all the memory devices simultaneously. Many timing problems can be eliminated by synchronous operation, using memory devices which operate on the *master-slave* principle.

There are several different types of counter. Examples are *up-counters*, *down-counters* and *modulo-n counters*. In all cases they can be designed using a next-state ROM and clocked register. Counters are almost invariably *preset* to an *initial state*. The initial state of a *preset counter* can be selected by the user.

Up-down counters or *reversible* counters are examples of *conditional* sequential logic circuits. The counter can be made to count 'up' or 'down' by the application of a *direction control input*. A fully synchronous up-down counter is arranged to ensure that the control input can only take effect on the receipt of a clock pulse. *Divide-by-n* counters are used extensively for *frequency division* in modern electronic systems. Counters can be designed to produce a *carry output* for the duration of the final state in the count sequence. Several modulo-10 counters can be connected in *cascade*, using the carry output, in order to count over a number of decades and avoid the use of a large capacity ROM.

The *programmable counter* is commonly available in integrated circuit form. It provides a good example of how a preset number can be loaded into a practical device to produce a counter with a desired modulo number.

Problems
for
Chapter 3

1. How would you set up a *D*-type flip-flop in order to register the occurrence of a single clock pulse?

2. A *D*-type flip-flop is connected as shown in the margin. The flip-flop is initially cleared to $Q = 0$. Draw the output waveform that would be generated by a sequence of pulses applied to the clock input.

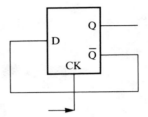

3. Describe how you would connect four *D*-type flip-flops to operate as a 4-bit clocked register. Draw a block diagram and label the input and output lines. Arrange for the clock to be applied to all four flip-flops simultaneously.

4. Indicate how you would use the *J–K* flip-flop as a memory element to store a single data bit on the receipt of a clock pulse.

   (Hint: The input is applied to a single line, while the flip-flop has two inputs. You must devise a simple combinational circuit to derive two inputs for the flip-flop).

5. The control of traffic lights gives an everyday example of sequence generation. Consider the basic problem of two sets of lights at some road works, with lights *R1*, *A1* and *G1* at one end and *R2*, *A2* and *G2* at the other.

   In logic terms the problem can be stated as the following basic sequence which is allowed to repeat as often as required:

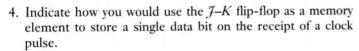

Stage	R1	A1	G1	R2	A2	G2
0	1	0	0	0	0	1
1	1	1	0	0	1	0
2	0	0	1	1	0	0
3	0	1	0	1	1	0.

   Show how the sequence could be implemented using a binary counter and output ROM assuming that the stages have equal duration.

6. Derive a next state ROM truth table for a modulo-8 natural binary down-counter which starts its count with an initial state of 111.

85

7. A 6 decade counter displays the following count:

   Decade: 10^5 10^4 10^3 10^2 10^1 10^0
   　　　　 9　 0　 9　 9　 1　 9.

   Which decades will change state on the receipt of the next clock pulse?

8. What preset numbers would be required to produce modulo-7, modulo-10 and modulo-12 operation from a 4-bit programmable counter?

9. Show in block diagram form how you would connect two 4-bit programmable counters to operate as a modulo-197 synchronous counter.

10. Design a modulo-6 up-counter using a 3-bit clocked register and next state ROM that can be used as a divide by 6 circuit. The output of this circuit is required to produce a square wave.

CHAPTER 4

Analogue/Digital Conversion

Introduction

Up until now we have concentrated on describing purely digital circuits. However, there is a large number of electronic circuits which employ purely analogue techniques and a large number of circuits which have both analogue and digital elements. Systems which employ both digital and analogue subsystems are frequently encountered in the fields of measurement and control. For example,next time you go into a large supermarket take a look at the weighing scales; you will find that the mechanical models are being, or already have been, replaced by electronic ones. The weight of your purchase is measured using an analogue transducer, which converts the weight into an equivalent analogue electrical signal, and this signal is converted into a digital form for display and price calculation. The multiplication needed to determine the price is easier and more accurate when implemented in a digital system.

The disc-recording industry is another example of the widespread adoption of combined analogue and digital systems. For the last few years the major recording studios have been producing all master tapes in a digital form, perhaps ready for the day that hi-fi goes digital. The next generation of family saloon cars will also use combined analogue and digital systems, to control exhaust emission, combustion timing and anti-skid braking systems, as well as to present a numerical display of speed, fuel consumption, temperature and so on.

However, before such systems can be designed it is first necessary to learn how to combine, or interface, analogue and digital subsystems. Fortunately, only a few basic building blocks are required, and they are the subject of this chapter. We have not attempted to examine the full range of manufactured devices available for interfacing, but only to select a few examples that illustrate the basic function and operation of each type.

Digital to Analogue Conversion

Binary weighted digital-to-analogue converter

digital to analogue converter

The first interface device is the *digital-to-analogue (D/A) converter*. Its function is to convert a digital input codeword into an analogue output signal, and so it forms the output interface from a digital to analogue subsystem. Figure 66 shows a simplified circuit diagram for such a converter. The four bit register stores the digital codeword to be converted. The outputs of this register are fed to the inputs of a summing amplifier of the type described in the Appendix. Notice that there is an important difference, which provides the key to the successful conversion from binary to analogue. In Figure 66 the summing resistors do not have equal values. Indeed they have values in the ratio $1:2:4:8$ and they are

binary weighted resistor network

termed a *binary weighted resistor network* because they have the same weighting as the digits in a binary codeword. Remember that for a four digit binary codeword the least significant bit had the value $2^0 = 1$, the next $2^1 = 2$, the next $2^2 = 4$ and the most significant bit $2^3 = 8$.

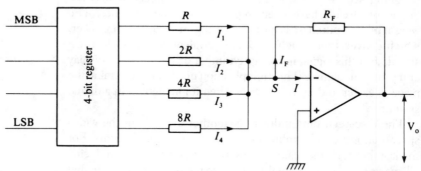

Figure 66   The buffered resistor network

However, notice that the least significant bit is connected to the largest resistor and the most significant to the smallest. The reason is contained in the analysis in the Appendix. Remember that due to the very high input impedance of the amplifier, all the current flowing out of the weighting network flows through the feedback resistor R_F. The node S is usually referred to as the summing junction, or virtual earth point. The output voltage V_O is given by the expression.

$$V_0 = -I_F \times R_F = -(I_1 + I_2 + I_3 + I_4) \times R_F$$

This being the case the most significant bit should produce the largest change in output voltage, implying the largest current change. Now since the binary digits are represented by the same voltage levels within the register independent of weighting, then this larger current is produced by connecting the most significant bit to the smallest resistor.

Example: What is the output voltage of the circuit of Figure 66 for an input codeword 1010, if a logic $1 = 5$ V and $0 = 0$ V, and $R = R_F = 1$ kΩ?

From the summing amplifier equation in Appendix B, the amplifier output will be

$$V_0 = I_F \times R_F = -V_S \times (\frac{1}{R} + \frac{0}{2R} + \frac{1}{4R} + \frac{0}{8R}) \times R_F$$

$$V_0 = -5 \times (\frac{1}{1 \times 10^3} + \frac{1}{4 \times 10^3}) \times 1 \times 10^3$$

$$V_0 = -6.25 \text{ V}.$$

So a binary codeword representing denary 10 gives a voltage of -6.25V. Similarly a codeword of 1100 representing denary 12 would produce a voltage of -7.5 V . The negative voltage is a result of using the inverting input to the amplifier and can be reversed by a subsequent inverting amplifier.

It is possible to use the D/A converter of Figure 66 to generate a variable d.c. output voltage. If, for example, the register were to be replaced by a 4-bit binary counter, the output voltage would increase in a series of steps as the counter is incremented. Once varying signals are generated, it becomes necessary to take into account the dynamic characteristics of the circuit. Of particular importance is the time taken for the output voltage of the op-amp to settle to a steady-state value after a change in the input codeword. This time interval is termed the *settling time.*

settling time

The settling time of a D/A converter is controlled by two factors. The first of these is the amount of current flowing in the resistor network that is available to charge any stray circuit capacitance. The second is the time response of the op-amp to a step input. Both these factors have a maximum influence when the most significant bit of the codeword changes state, because this bit contributes the most current and hence causes the greatest output voltage swing. The time taken for a most significant bit transition is therefore normally quoted by manufacturers as the D/A converter settling time.

D/A converter with a stabilized voltage source

The major problem with the circuit of Figure 66 is that the logic gate output voltages are not well defined. They are only specified to lie within a certain range of voltages. The accuracy of the D/A converter can be improved by deriving the voltage levels from a stabilized voltage source. In such a design the logic signals are used to switch the resistors between the voltage source and common as shown in Figure 67. Apart from this modification, the circuit operates as before. While the introduction of switches removes the problem associated with variable voltage sources, their use may introduce other errors, particularly if solid-state switches are employed to maximize switching speeds.

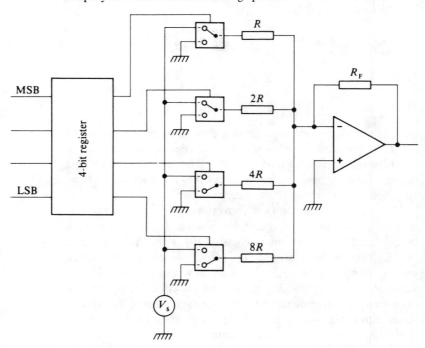

Figure 67   Binary weighted D/A converter with a stabilized reference voltage

These solid-state switches can be of the p-channel JFET type. Their circuit symbol is as shown in Figure 68. To use the JFET as a switch, the input signal is applied to the source and the output signal is taken from the drain of the device. An appropriate control signal must be applied to the gate to open and close the switch. When the gate is held at 0 V the source-drain channel conducts with a low, but non-zero, resistance, typically 300 Ω for the E177 JFET. This corresponds to the 'on' state of the switch. If the gate is held at $+5$ V the channel resistance increases to around 10^{10} Ω, and only a small leakage current flows from source to drain, typically 1 nA for the E177 device. This corresponds to the 'off' state of the switch. Because the control signal is either 0 V or 5 V we can use any of the commonly available logic devices to control the state of the switch. But note that if we follow a positive logic convention, that is 5 V = logic 1 = switch on, we must provide an extra stage of inversion between the gate terminal and the control signal, otherwise a logic 1 input will turn the switch off.

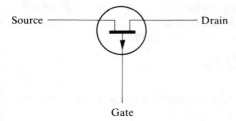

Source ——————————— Drain

Gate

Figure 68   The circuit symbol for a p-channel JFET

Rather that add to circuit complexity by explicitly showing this extra inversion, always assume that the switching device incorporates such circuitry and is therefore 'on' when a logic 1 is applied to the control terminal, and is 'off' for a logic 0. Now let us determine what effect the introduction of the switches has upon the performance of the D/A converter. To simplify the analysis we will employ two low-frequency circuit models for the switch. Using these models an 'on' state switch is replaced by a resistor R_{on} which represents the resistance of the source-drain channel, and an 'off' state switch is replaced by a current source I_L which represents the leakage current.

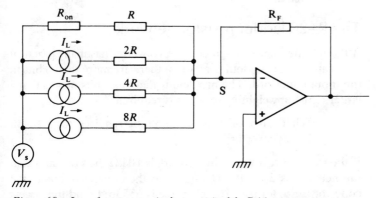

Figure 69   Low-frequency equivalent circuit of the D/A converter

Figure 69 shows how we can use these models to establish the errors in the D/A converter of Figure 67 when the input codeword is 1000. For this example the switch in series with resistor R is replaced by resistor R_{on}, and the other three switches are replaced by current sources I_L. The total current $_s$ entering the summing junction is

$$I_s = \frac{V_s}{R_{on} + R} + 3 \times I_L$$

(You will see from Appendix A that an ideal current source has an infinite source resistance and acts as an open circuit to other sources, while an ideal voltage source has zero source resistance and so behaves as a short circuit to other sources.) If the switches were ideal (that is, no 'on' resistance or leakage current) the current I entering the summing junction would be

$$I = \frac{V_s}{R.}$$

The difference $(I-I_s)$ times the feedback resistor equals the error in the output voltage of the op-amp.

For most practical applications we can ignore the leakage current of an analogue switch, because its contribution to the total error is very small. The significant part of the error arises because the switch has a finite 'on' resistance which increases the output resistance of the resistor network, and so reduces the gain of the amplifier. We can minimize the error by increasing the value of R. You may think that increasing the feedback resistor R_F could achieve the same results by increasing the gain to allow for the reduction caused by switch resistance. However, increasing the feedback resistor is not an optimum solution, because the output voltage error depends upon the number of switches in the 'on' state, and so varies with the input codeword, while the feedback resistor is fixed.

The R–2R ladder resistor network

Although the binary-weighted D/A converter has many practical applications, it is seldom used when there are more than 6 bits in the input codeword. The problem is that more bits means that a larger range of weighting resistors is required.

What range of resistors would be required for a 16-bit D/A converter?

If the smallest resistor in the network is $10 \text{ k}\Omega$ then the largest in the network is $2^n \times 10 \text{ k}\Omega$ where n is the number of bits in the codeword. So, for $n = 16$, the range is 65536:1, and the largest resistor is 655 MΩ. It is very difficult to manufacture such large resistors which maintain the appropriate weighting accurately, either as discrete components, or on an integrated circuit.

An alternative resistor network exists which incorporates resistance values that only vary over a 2:1 range for any number of bits in the codeword. The principle involved is illustrated in Figure 70. The current I entering node N must leave by way of the two resistors R_1 and R_2 (Kirchhoff's current rule). If these resistors are equal, $I/2$ will flow in each branch. Such a resistor network gives us a simple method for dividing the current flowing in a network. The principle can be extended indefinitely; the only requirement is that the equivalent resistance of each current path leading away from the node is the same. Figure 71(a) shows a circuit for dividing the input current by 4. $I_2 = I/4$, but to check this we must first calculate I_1. This is done by calculating the equivalent resistance of the circuit to the right of line A, and then applying Kirchhoff's rule. The equivalent resistance R_{eq} is equal to R plus $2R$ in parallel with $2R$, that is

$$R_{eq} = R + \frac{2R \times 2R}{2R + 2R}$$

$$R_{eq} = 2R.$$

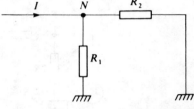

Figure 70   Current division at a node

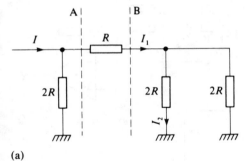

(a)

Figure 71   Current dividing networks

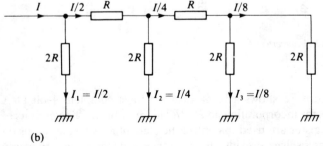

(b)

I_1 is therefore equal to $I/2$. Because the resistances of the two branches to the right of line B are equal, $I_2 = I_1/2 = I/4$. The circuit of Figure 71(b) extends the idea one stage further to give a current $I_3 = I/8$, as well as $I_2 = I/4$ and $I_1 = I/2$. Notice that all the resistors are either R or $2R$, hence such networks are referred to as *R–2R ladder* networks.

Now let us see how we can use the *R–2R* ladder to build a D/A converter. Our problem is to ensure that the current controlled by each bit of the input codeword is related to a bit's position in the codeword. If, for example, the most significant bit of a natural binary codeword switches in a current I, then the next most significant bit must switch in a current $I/2$, and so on. Remember that in the case of the binary-weighted network the current switch by each bit was determined by a resistor, so the resistance was increased by a factor of 2 going from each bit to the next less significant bit. If we use the *R–2R* ladder the current switched by each bit is determined by the point along the ladder at which it is tapped off.

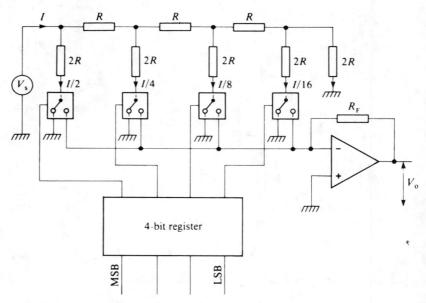

Figure 72   An R-2R ladder network D/A converter

Figure 72 shows a block diagram for a complete 4-bit D/A converter incorporating an *R–2R* ladder network. The outputs of the register are used to control the state of a switch, and thus do not themselves provide the reference voltages for the network. The op-amp operates in exactly the same way as for the binary network D/A converter, by summing the current flowing into its inverting input.

What is the full-scale voltage output of an 8-bit D/A converter designed around a $R–2R$ ladder network and a 5 V reference source, if the feedback resistor is $10\,k\Omega$ and $R = 5\,k\Omega$?

The output voltage of the D/A converter is given by

$$V_O = -I_F \times R_F.$$

To calculate I_F we need to know the current flowing in each switch, but first we must determine the maximum current flowing into the ladder. This current is equal to the reference voltage divided by the $R–2R$ ladder input impedance. Combining all the parallel and series resistors of the ladder gives an input impedance of R, so the maximum current is $2\,V/5\,k\Omega = 1\,mA$. The maximum output voltage is obtained for an input codeword consisting entirely of logic 1's, so the current entering the feedback loop is

$$I_F = \frac{I}{2} + \frac{I}{4} + \frac{I}{8} + \frac{I}{16} + \frac{I}{32} + \frac{I}{64} + \frac{I}{128} + \frac{I}{256}$$

$$I_F = \frac{255}{256}\,I.$$

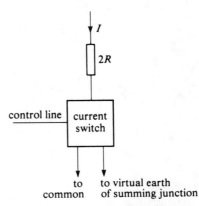

Hence the output voltage is

$$V_O = \frac{-255}{256}\ 1\,mA \times 10\,k\Omega$$

$$V_O = -9.96\,V.$$

Figure 73   A current switch for the R-2R ladder

Another important design improvement to the D/A converter of Figure 72 compared with that of Figure 67 is the switching arrangement. In this example, the current is switched between the summing-node of the op-amp and common as shown in Figure 73. Because these two points are at approximately the same potential, there is no appreciable voltage change in the circuit, hence the speed limitations imposed by stray circuit capacitance are minimized, because capacitors do not charge up, and so on. The D/A converter settling time is now limited only by switching speed and the op-amp response.

Bipolar conversion

In all of the D/A converter designs we have examined so far the output voltage has been negative. This is because the reference voltage was chosen to be positive and the D/A converters have employed inverting amplifiers. We could obtain a positive output voltage from the same D/A converter by using a negative reference voltage. So by choosing the appropriate polarity for the reference voltage we can build a D/A converter that will produce either polarity of output, but not both. Such single-polarity output D/A converters are called *unipolar converters.*

unipolar converters

bipolar converters

In many practical applications, it is necessary to design a D/A converter that will produce both positive and negative outputs; they are called *bipolar converters.* To obtain bipolar outputs it is necessary to modify the design of the unipolar D/A converter so that it will accept binary input codewords which contain both polarity and magnitude information. Figure 74(a) shows the output voltage for a unipolar converter. Remember that the polarity of the output only depends upon the polarity of the reference voltage. In this example we have chosen a negative reference resulting in a positive output from the amplifier. The smallest output is produced by a binary input of 0000, the largest by an input of 1111. Figure 74(b) shows the simplest method of producing a bipolar output. In this case a voltage equal to half the reference voltage V_{ref} has been subtracted from the output of the D/A converter.

Let us now calculate the full range of the D/A converter output voltage for the case $R_F = R$. For an input codeword of 0000 the D/A converter output is 0 V, but from this we must subtract $V_{ref}/2$, so the actual output is $-V_{ref}/2$. When the input codeword is 1111, the D/A converter output is $15/16\ V_{ref}$, so the offset output is $+7/16\ V_{ref}$. The full-range output of the bipolar D/A converter is therefore $+7/16\ V_{ref}$ to $-V_{ref}/2$. For this modified converter an input of 0000 gives the maximum negative output while 1111 gives the maximum positive output. A zero output is achieved when a binary input of 1000 is applied to the converter. This type of coding is referred to as *offset binary,* simply because it is a normal binary D/A converter with the output offset by half the full-scale output.

offset binary

There are several methods for actually producing the necessary offset voltage, but by far the most popular is to inject an equivalent offset current, I_{OFF}, into the summing junction of the op-amp as shown in Figure 75. The value of this current is adjusted to be equal in magnitude, but opposite in sign to the current injected by the ladder network when a binary input of 1000 is applied to the converter. This equal and opposite current ensures that no current flows through the feedback resistor R_F for an output codeword of

96

1000 resulting in zero volts output from the amplifier.

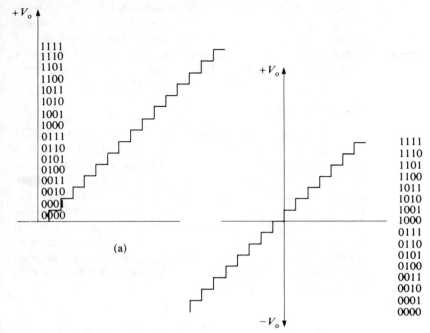

Figure 74   Unipolar and bipolar converter outputs

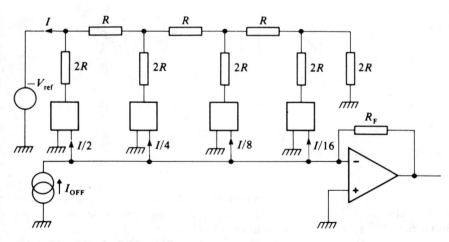

Figure 75   A bipolar D/A converter

In practice, it is very difficult to preserve the equality between the offset current and the ladder network current produced by an input of 1000 over an extended period of time. Component aging

97

and temperature variations inevitably introduce a small difference between these two currents, which results in some current flowing through the feedback resistor and producing a non-zero output voltage. The difference between the actual output voltage and zero volts is called the *bipolar offset error,* and for a high-performance D/A converter this error should be no greater than the output voltage obtained from the least significant bit of the input codeword.

bipolar offset error

Quantization

So far we have shown you how a D/A converter can be used to generate a d.c. voltage, or, when driven from a binary counter, a staircase waveform. In practice, D/A converters can be used to generate any waveform given the appropriate input codewords. However, using the discrete output levels of a D/A converter to represent an analogue waveform involves the process of quantization. For example, suppose we wanted to generate the sawtooth waveform shown in Figure 76(a) for use in the timebase of an oscilloscope. As a first approximation we could try the staircase waveform obtained by driving a D/A converter from a 4-bit counter, shown in Figure 76(b).

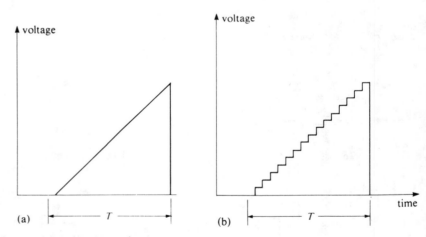

Figure 76   Analogue and digital ramp waveforms

quantization levels
quantization interval

The 16 possible values of the D/A converter output voltage are called the *quantization levels,* and the difference in voltage between two adjacent quantization levels is termed a *quantization interval.* The quantization interval is also equal to the output generated by the least significant bit of the binary input codeword. The slope of the staircase waveform is determined by the rate at which the input codewords to the D/A converter change. For example, to obtain a steep slope we must change the D/A converter input codewords quickly, while for a shallow slope they need only be changed slowly. If we want a uniform slope, the duration of each step is equal to the period of the ramp divided by the number of quantization levels. In our example the period of the ramp is T, so the duration of each step is $T/16$, say τ, then τ is the period of the clock pulse applied to the counter. Remember that τ must be long enough to allow the D/A converter output to settle before the arrival of the next clock pulse.

We suggested the staircase waveform as a first approximation to the ramp, but how good an approximation is it? Figure 77(a) shows a short portion of the two waveforms of Figure 76 drawn to a much larger scale. You can see that during the interval t_1 to t_2 the required ramp voltage increases from V_A to V_B, while the D/A converter output is constant at V_1. The two waveforms are therefore equal at only one instant of time during the interval t_1–t_2, and at all other points they differ. This is due to the fact that the smallest

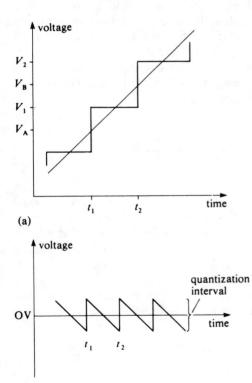

(a)

(b)

Figure 77   Quantization errors

increment in the D/A converter output is equal to the contribution of the least significant bit; that is, one quantization interval. In terms of our approximation this difference produces an error, known as the *quantization error.*

quantization error

We can see the effects of the quantization error by subtracting the ideal ramp voltage from the D/A converter output voltage which approximates to it. For example, using the waveforms of Figure 77(a), the difference between the D/A converter output and the ramp is shown in Figure 77(b). You can see that the error alternates about zero volts, and has a peak-to-peak value equal to the quantization interval. It is convenient to model the quantization error illustrated in Figure 77(b), as noise that is added to the ramp voltage to produce the final D/A converter output. Because this noise is due to quantization, it is given the name *quantization noise.*

quantization noise

To reduce the quantization error we must use a D/A converter controlled by more bits because if the full-scale output voltage remains the same this will reduce the size of the quantization interval. For example, the quantization error for a 4-bit converter is always 1 part in 15 (full-scale output is equal to 15 quantization intervals), and is 1 part if 225 for an 8-bit D/A converter, no matter what the full-scale output voltage.

dynamic range

An alternative way of specifying the quantization effect is in terms of the *dynamic range* of the D/A converter. The dynamic range (dr) in decibels is defined as

$$dr = 20 \log_{10} \frac{\text{full-scale output}}{\text{quantization interval}}$$

and indicates the difference between the smallest non-zero and largest output voltages that can be generated by a D/A converter. The dynamic range can also be obtained by substituting into the above equation the actual value for the full-scale output in terms of quantization intervals. For an n-bit converter there are 2^n quantization levels, but only $2^n - 1$ quantization intervals, hence the full-scale output is $(2^n - 1) \times q$, where q is the quantization interval. The dynamic range is therefore

$$dr = 20 \log_{10} \frac{(2^n - 1) \times q}{1 \times q}$$

$$dr = 20 \log_{10} (2^n - 1)$$

A simple rule of thumb to remember is that each bit of the input codeword of a D/A converter adds 6 dB to the dynamic range.

100

Summary

Digital-to-analogue converters are devices that convert a digital codeword into an analogue voltage by means of resistor weighting networks. In particular, we examined the binary weighted resistor network and the $R-2R$ ladder network.

We saw that accurate D/A converters require a stabilized reference voltage source, and this in turn requires the use of analogue switches. Although the switches introduce their own forms of error, these can be minimized by careful design. Switching current rather than voltage reduced the settling time of the resistor network, and so speeded up the conversion time. The digital generation of waveforms involves the process of quantization, and introduces quantization error. This quantization error can only be reduced by increasing the number of bits in the converter.

Analogue-to-digital converters

Devices which convert analogue voltage or current into digital codes are called analogue-to-digital (A/D) converters. The simplest type of conversion occurs when the input signal is a positive d.c. voltage and this is to be converted into a binary codeword. So let us begin with this situation and examine one possible type of converter.

The counter ramp A/D converter

A block diagram of this type of converter is shown in Figure 78. It comprises a D/A converter, a binary counter, a comparator and a clock plus associated control logic. The comparator may be a new device to you and is described more fully in Appendix B, but for the moment a brief description of its operation will suffice. It is a

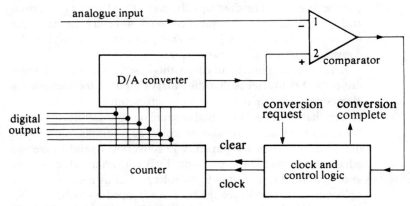

Figure 78   A counter-ramp A/D converter

two-input, one-output device, and is used to determine which of the two input voltages is the larger. The output signal is made to be compatible with logic circuits and is logic 0 if one input is larger and logic 1 if the other is larger. (A circuit diagram for such a device is shown in Appendix B.)

The A/D converter works as follows, when a conversion is required a signal is sent to the converter from the digital subsystem requesting a conversion. This signal clears all the counter outputs to 0, and hence forces the D/A converter output to zero volts. Input 2 of the comparator is therefore also zero volts, so that if the analogue signal at input 1 of the comparator is a positive voltage the comparator output is set to logic 0. Once the D/A converter output has settled, a clock increments the counter, thus increasing the D/A converter output. After each clock pulse has incremented the counter, the D/A converter output changes to a new value. If this new output is greater than the input signal, the output of the comparator changes to a logic 1, and this stops the clock signal.

If, on the other hand, the D/A converter output is less than the analogue input, the counter is incremented by another clock pulse. This comparison and increment sequence is repeated until the D/A converter output exceeds the input voltage. Figure 79 is a timing diagram for the conversion process, and shows the inter-relationship between the conversion request signal, the clock pulses, the D/A converter output and the comparator output. Notice that the counter increments on the rising clock edge, and that the comparator output only changes when the D/A converter output exceeds the input voltage. This design of converter is called a counter-ramp A/D converter. The conversion cycle is completed once the comparator output changes from logic 0 to logic 1. At this point the 'clock and control logic' block generates an *end of conversion signal* to indicate to the digital subsystem that the binary representation of the input voltage can be read from the counter outputs. The time lapse between the start of a conversion and the generation of the end of conversion signal is known as the *conversion time* of the A/D converter, and the reciprocal of the conversion time is the *conversion rate*.

An important point to note from this description of the counter-ramp A/D converter is that the binary output of the converter is always equal to one of the D/A converter input codewords. This means that analogue-to-digital conversion is also subject to quantization errors and quantization noise. Figure 80(a) shows how quantization errors arise in an A/D converter. V_1 and V_2 are two adjacent quantization levels of the D/A converter output voltage and V_i is the d.c. analogue input voltage. Let us assume that the D/A converter output voltage has been stepped up to the level V_1. Because $V_i > V_1$, the comparator output has stayed at a logic 0. On

end of conversion signal

conversion time
conversion rate

102

the next clock pulse the D/A converter output rises to V_2, and because $V_2 > V_i$ the comparator output changes to a logic 1, and so stops the conversion cycle.

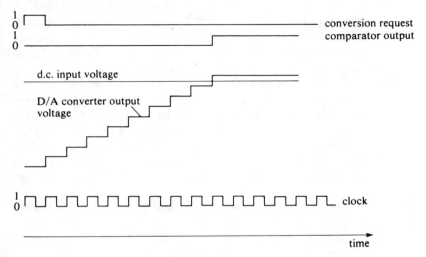

Figure 79   Timing diagram for the counter ramp A/D converter

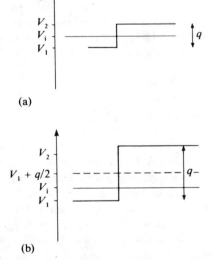

Figure 80   Quantization errors in an A/D conversion

However, the binary codeword read from the counter is an exact representation for V_2 not V_i, hence there is an error. The maximum error occurs when $V_i = V_1$ and is equal to the quantization interval q. It is possible to reduce this error to half a quantization interval by adding a voltage offset of $q/2$ to the D/A converter output. Let us see what effect this has on the conversion error.

103

When the D/A converter output equals V_1, the actual comparator inputs are V_i and $V_1 + q/2$. From Figure 80(b) we can see that $V_1 + q/2 > V_i$ so the comparator output changes state and the conversion process stops. The codeword at the input of the D/A converter represents V_1, not $V_1 + q/2$; because the $q/2$ offset is added to the converter's output. Hence the digital value assigned to V_i equals V_1, and the conversion error is $|V_i - V_1|$ which is less than $q/2$.

What happens when $V_1 + q/2 > V_i > V_2$? In this case, the comparator inputs must be V_i and $V_2 + q/2$ for the conversion to stop. But the input codeword to the converter represents the voltage V_2, so the conversion error is $|V_i - V_2|$ which is also less than $q/2$. Adding an offset of $q/2$ to the D/A converter output is analogous to rounding in everyday arithmetic, for example, we might round 101.7 upwards to 102 or round 101.2 downwards to 101.

The advantage of a reduction in quantization error, and hence quantization noise, usually outweighs any increase in A/D converter circuit complexity, and so rounding has been adopted almost universally in A/D converter designs.

The one major drawback of the counter-ramp A/D converter is its relatively long conversion time for large input signals, this arises from the fact that the D/A converter output must ramp from zero. The conversion time can be reduced by decreasing the counter clock period, but there are practical limitations to this approach, because the clock period must allow the D/A converter output to settle after each counter increment.

The successive approximation A/D converter

We would now like to view another A/D converter design; one which allows much faster conversion rates than the counter-ramp type. It is called a successive-approximation converter. Figure 81 is a block diagram for a 4-bit successive approximation A/D converter. The major difference from the counter ramp A/D converter is that the counter has been replaced by a 4-bit register. The states of the output bits of this register are controlled by the 'clock and control logic' block.

The converter works as follows. At the start of a conversion the first clock pulse clears all the register outputs to logic 0. On the second clock pulse the most significant bit of the register is set to logic 1, which causes the D/A converter output to change to half its maximum value. If the analogue input is less than D/A converter output, the comparator output causes the logic control to leave the register bit set to a logic 1. On the next clock pulse this process is repeated for the next most significant bit, and so on until all four bits of the register have been tested. The total process is somewhat like a guessing game in which we think of a number and you try to guess what it is. After each of your guesses we will only tell you whether our number is less than or greater than your guess. Eventually you will get the right answer.

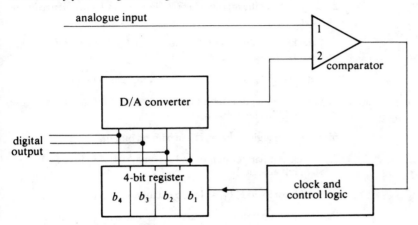

Figure 81   A successive approximation A/C converter

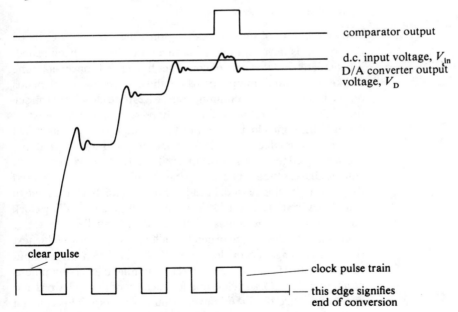

Figure 82   Timing diagram for the successive approximation A/D converter

Let us work through all the steps for the 4-bit converter, for a particular d.c. input voltage. Figure 82 is a timing diagram for the

105

conversion process, and it illustrates the interrelationships between the clock pulse train, the D/A converter output, and the comparator output. V_{in} is the d.c. analogue input, and V_D is the D/A converter output. There are ten stages to the conversion as follows:

1. Clear all register outputs to logic 0. This forces $V_D = 0$ and the comparator low.

2. Set most significant bit to logic 1. This forces V_D to half maximum value.

3. Check comparator output and re-set bit b_4 to 0 if $V_D > V_{in}$. Because $V_D < V_{in}$ this bit remains set.

4. Set next bit b_3 to a logic 1, $V_D = 3/4$ maximum output.

5. Check comparator output and re-set b_3 to 0 if $V_D > V_{in}$. Because $V_D < V_{in}$ this bit remains set.

6. Set b_2 to a logic 1, $V_D = 7/8$ maximum output.

7. Check comparator output and re-set $b_2 = 0$ if $V_D < V_{in} \cdot V_D > V_{in}$ so b_2 left at logic 1.

8. Set b_1 to a logic 1, $V_D = 15/16$ maximum output.

9. Check comparator output and re-set b_1 to 0 if $V_D > V_{i}$, b_1 is re-set because $V_D > V_{in}$.

10. Signal conversion completed.

It is a feature of the successive approximation converter that a conversion is never completed until all the bits have been tested. This means that this type of converter has a fixed conversion time unlike the counter-ramp type where the conversion time depends on the input voltage. Examining Figure 82 more closely we can see that the setting, testing and possibly re-setting of a bit must all occur within one clock cycle. For this example, we have assumed that the rising edge of the clock signal is used to set a bit and that the falling edge must not occur until the D/A converter output voltage has settled. The comparator output must be ignored during this settling period, because its output might change due to transient variations in the D/A converter output. The falling clock edge therefore determines when the data bit will be re-set, if necessary. The period between the falling clock edge and the next rising edge is again provided to enable the D/A converter output to settle if the tested bit was re-set. The total conversion time is equal to $n + 1$ clock cycles, one for each bit of the codeword and one to initialize the D/A converter output to zero. This is much faster than the counter-ramp A/D converter, if we ignore the few occasions when the input voltage is near zero. An additional advantage is that the conversion time is independent of input amplitude, which is useful in systems that require a constant conversion rate.

> What is the maximum conversion rate that can be guaranteed with a n bit counter-ramp converter?

The allowed conversion time must be equal to 2^n clock cycles to allow for the case of maximum input voltage. This gives a conversion rate of

$(\frac{1}{2^n}\times$ clock frequency) conversions per second.

The conversion of a.c. signals

The finite conversion time of an A/D converter presents no special problems when converting d.c. signals, but what about a.c. signals which will change in voltage between successive conversions? To see how to convert a.c. signals let us examine a simple analogy, that of plotting a graph of room temperature against time. It is not practicable to measure the temperature at every instant of time; it would take several seconds just to read a thermometer and write down the result. Instead measurements are taken at fixed intervals of time, and the result is a sample of all the possible temperature measurements. The rate at which measurements are taken, or sampled, is the *sampling rate*, and this determines the eventual agreement between the graph and the actual temperature variations.

sampling rate

The faster the sampling rate, the better the graph will represent the actual temperature but simply sampling as fast as possible is sometimes inefficient. Figure 83 shows three samples of the room temperature, A_1, A_2 and A_3, taken at times t_1, t_2 and t_3 respectively. If the actual room temperature variation between times t_1 and t_3 can be approximated by a straight line, then the measurement at t_2 is superfluous, because it could have been predicted from A_1 and A_2, using linear interpolation. On the other hand, if the actual temperature variation between t_1 and t_3 is very rapid, as shown in Figure 83(b) then linear interpolation will not enable accurate prediction of the temperature between t_1 and t_3. To maintain accurate prediction more sample points would be needed.

This argument would suggest that the sampling rate should be related to the rate of change of temperature, which in turn will be related to the frequency content of the temperature waveform. Indeed there is a simple rule that relates the minimum sampling rate to the signal bandwidth, it is called the *sampling rule*. We do not intend to prove this rule, but will merely state it. The sampling rule states that a continuous signal with a bandwidth from d.c. to f Hz can be completely represented by, and reconstructed from, a set of equally spaced samples of instantaneous voltage which are taken at a rate which exceeds $2f$ samples per second. So for an instrumentation signal with a bandwidth of d.c.–10 kHz, samples must be taken at a minimum rate of 20 000 samples per second. Whenever possible the sampling rate should be chosen to be adequate for the bandwidth of the signal, in which case the sampling process is straightforward. Occasionally though, the situation arises where the available sampling rate is inadequate for the bandwidth of the signal to be measured. In this case some of the information contained in the signal must be discarded by limiting the signal bandwidth. Let us show you why.

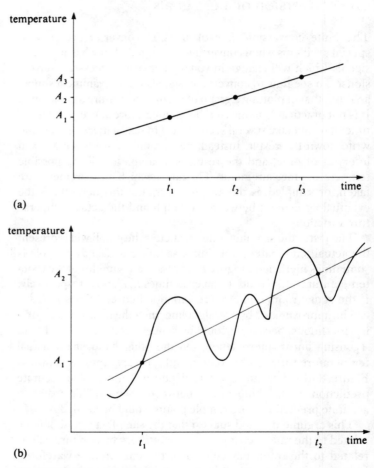

(a)

(b)

Figure 83   Sampling a.c. signals

Figure 84(a) shows a sine wave S_1 that has been sampled at a rate below that required by the sampling rule giving the sample points shown. Figure 84(b) shows that there is a second lower

108

frequency sine wave S_2 that is better represented by the same set of samples. S_2 is called the alias of S_1. Given only the sample points, it is impossible to determine whether they are the result of sampling S_1 or its alias S_2. Whenever we sample a signal too slowly it is possible that energy contributed by high-frequency components can be attributed to their low-frequency aliases. The errors that arise are said to be caused by *aliasing*.

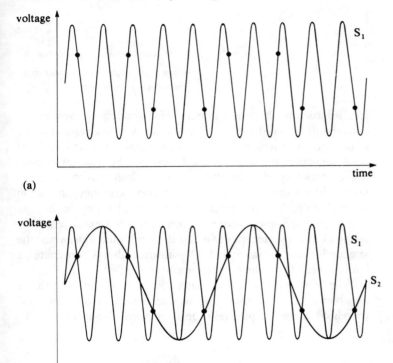

(a)

(b)

Figure 84   The effects of under sampling

High-frequency noise also creates errors in the conversion process. For example, Figure 85 shows that the true sample voltage at time t_1 is V_1, but due to high-frequency noise the measured value becomes V_2. To minimize errors due to both aliasing and high-frequency noise it is essential that the A/D converter is preceded by a low-pass filter with a bandwidth determined by the available sampling rate. Such filters are frequently referred to as *anti-aliasing filters*.

The sampling rule tells us at what rate to make conversions, but there is still another problem associated with a.c. signals. Figure 86 shows part of a sine wave to be converted by an A/D converter.

109

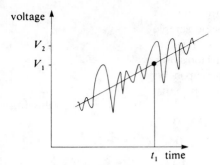

Figure 85   Sampling errors caused by high-frequency noise

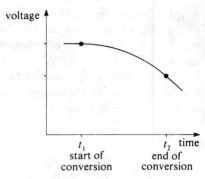

Figure 86   Variation of input signal during a conversion period

Let us assume that the sampling rate is adequate and that a conversion starts at time t_1. Because an A/D converter takes a finite time to complete a conversion, the end of conversion signal does not occur until time t_2. A problem arises because throughout the conversion period the input signal is changing in amplitude. If the maximum variation of the input signal exceeds one quantization interval of the A/D converter the effects are alarming. In the case of a counter-ramp converter the conversion may never terminate as the D/A converter tries to follow the input variations, whilst the successive-approximation A/D converter always terminates a conversion, but produces an erroneous result.

What we need is a device to precede the A/D converter that is capable of taking an instantaneous sample of the input signal and then holding the sampled value until the conversion is completed.

Sample-and-hold devices

We previously described the need for a device that instantaneously samples an input waveform upon command, and holds it long enough for an A/D converter to complete the conversion cycle. Such a device is called a 'sample-and-hold'.

A simple circuit that meets the basic requirements is illustrated in Figure 87. When the switch, for example an E177 JFET, is in the closed position the voltage across the capacitor equals the input voltage. Upon command the switch is opened and the charged capacitor retains the input voltage. However, practical versions of this simple circuit are often inadequate for use in precision measurement systems for several reasons.

First let us consider the input of the circuit. Connecting a capacitor across the signal source, as with the switch closed, will load the circuit producing the signal. This loading can be overcome by preceding the switch with a unity gain buffer amplifier which has a high input impedance and low output impedance. Such a buffer amplifier also ensures that on subsequent closures of the switch the time constant for charging the capacitor is limited by the switch 'on' resistance, and not by the preceding circuit. Now let us look at the output of the sample-and-hold, bearing in mind that it is connected to an A/D converter with a finite input impedance. Any current flowing out of the capacitor when the switch is open will reduce the stored charge level and hence the voltage across the capacitor. If this voltage change is greater than a quantization interval, errors will arise in the conversion process. Adding a second high input impedance unity gain buffer amplifier after the capacitor will minimize the charge loss due to loading, by the A/D converter for example, but cannot eliminate it entirely. This is because a small leakage current will flow through the capacitor dielectric regardless of other losses. The total charge loss from the capacitor is specified in terms of the rate of change of voltage, dv/dt, and is called the *droop rate*. A typical block diagram for a complete sample-and-hold is shown in Figure 88.

droop rate

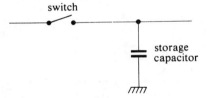

Figure 87   A basic sample-and-hold circuit

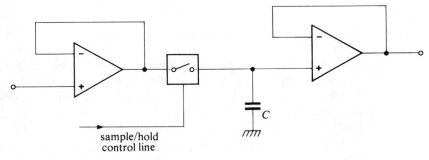

Figure 88   A buffered sample-and-hold

Specifying the sample-and-hold

Having now developed the circuit for a sample-and-hold, we want to examine the parameters that must be specified by the system designer. The first of these is related to the speed at which the switch can change from the 'closed' to the 'open' state upon receipt of a command to hold. Real switches cannot change state instantaneously and there is always a short delay between the inception of the hold command and the time the capacitor voltage ceases to follow the input. This delay is called the *aperture time*, and its effect is to introduce an error in the stored voltage level as shown in Figure 90. At the inception of the hold command the input voltage is v_1, but because the capacitor continues to track the input during the aperture time, the voltage actually stored is v_2. It is up to the designer to ensure that the error $v_2 - v_1$ is commensurate with the overall system accuracy, such as specifying a maximum error of half a quantization interval.

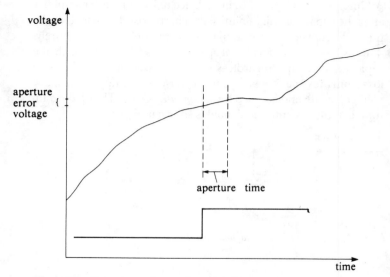

Figure 89   Error due to the sample-and-hold aperture time

There is also delay associated with the change in state from 'hold' to 'sample'. Upon receipt of a 'sample' command, the switch closes and the capacitor either charges or discharges until its voltage equals the input voltage. The switch 'on' resistance limits the current that can flow between the buffer amplifier and the capacitor and so limits the response time of the sample-and-hold.

The response time can be specified in terms of the time constant produced by the switch resistance and the storage capacitor. Manufacturers quote the response in terms of the time required for the capacitor voltage to settle to within a fixed percentage of its final value. We can estimate the response time with the aid of Figure 91, which shows the buffer amplifier output stage, modelled

112

as a voltage source and series output resistance, together with the switch 'on' resistance and the storage capacitor C. The time constant for charging C is $(R_s + R_{on}) \times C$, but since R_{on} is typically 300 Ω and R_s is usually less than 1 Ω, it is reasonable to approximate the time constant as $R_{on} C$. For the capacitor voltage to settle to 0.01 per cent requires roughly 9 times constants. The maximum settling time, and hence acquisition time, occurs when the capacitor voltage has to change full scale, say from -10 V to $+10$ V, so this is the condition normally quoted by the manufacturers.

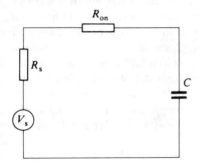

Figure 90   RC equivalent circuit of a sample-and-hold

The final parameter we want to consider is *feed through*, this tells us how much of the input signal appears at the output of the sample and hold when it is in the hold state. In all previous discussions of the errors associated with solid state switches we have assumed that an 'off' state switch can be modelled as a d.c. current source. For low-frequency signals this model is quite adequate, but it is not adequate for high-frequency signals.

This apparent discrepancy can be attributed to the effect of stray capacitance within the electronic switch. At low frequencies the stray capacitance offers a high impedance to the flow of alternating current through the source to drain channel, but at high frequencies the impedance decreases and so some of the input signal appears at the output. The fraction of the input signal to the sample and hold appearing at the output is termed the feed through.

feed through

Multiplexers

The A/D converters and sample-and-hold devices we described in the last two sections have all been single-input devices. Frequently though we want to sample several analogue signals, and convert each of the sample values to a binary codeword. The cost of providing an A/D converter and sample-and-hold for each analogue input can be prohibitively expensive and difficult to justify if the bandwidth of the input signals indicates a sampling rate below the maximum available from the A/D converter.

Converting several narrow bandwidth signals with a single fast A/D converter is a commonly used technique in modern interface subsystems, and is one example of time division multiplexing. Figure 91 illustrates how two input signals can be time-division multiplexed together using a rotary switch. As the switch rotates, so each input is connected in turn to the switch output. Figure 92 shows the two input signal waveforms and the switch output waveform. You can see that during interval t_1 input signal 1 appears at the output and during t_2 signal 2 appears at the output. To use time-division multiplexing in a multi-input analogue-to-digital conversion subsystem requires that the multiplexer is followed by a sample and hold and A/D subsystem as shown in Figure 93. Each signal has to be sampled fast enough to satisfy the sampling rule so the comparatively slow switching in Figure 91 will not do.

For two input signals each requiring a sampling rate of f_s samples/sec the switching should be at $2f_s$. For three signals $3f_s$ and so on. This enables each of the signals to be sampled at the correct rate.

What frequency is the sample and hold sampling the output of the multiplexer? Well, for one signal is would simply be f_s, for two signals $2f_s$, for three signals $3f_s$ and so on.

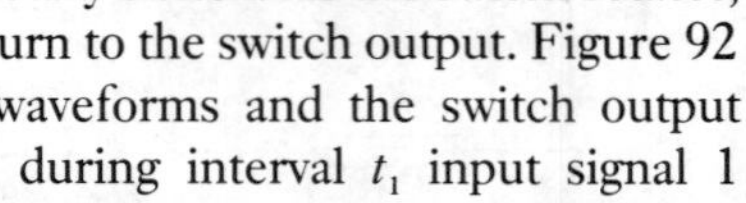

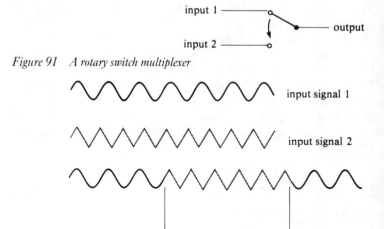

Figure 91   A rotary switch multiplexer

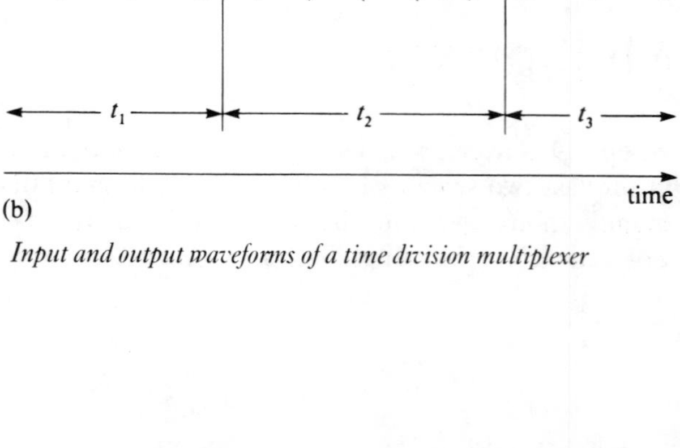

Figure 92   Input and output waveforms of a time division multiplexer

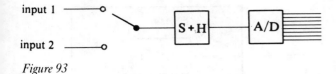

Figure 93

Solid-state multiplexers

Rather than a rotary switch, modern integrated-circuit multiplexers consist of a group of electronic solid-state switches together with all the necessary decoding logic to select the appropriate input, all in one package. Figure 94 is a schematic diagram of a four-input multiplexer. Note that only two control lines are necessary when the decoding is performed within the multiplexer, rather than using four lines, with one for each switch.

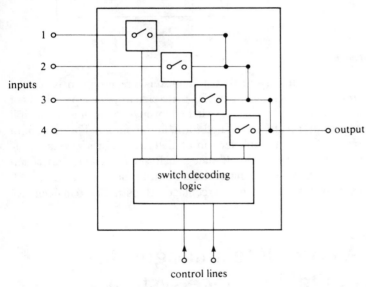

Figure 94   Schematic of a 4-input multiplexer

A second point to note is that there is no reference in the diagram to a common line. This is because this design assumes all the inputs and the output share the same common line. For this reason such multiplexers are called *single-ended* devices. Alternative designs are available that enable a single control line to select a pair of switches as shown in Figure 95. They are useful when the signal to be converted exists as the difference between a pair of conductors, so they are called *differential input* multiplexers.

single-ended

differential input

115

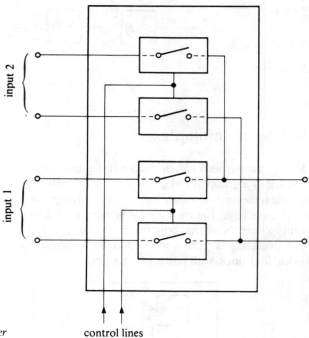

Figure 95   A differential input multiplexer   control lines

One of the errors that multiplexers are prone to is termed *cross-talk*. This refers to the fact that a signal applied to the input of a multiplexer which is not switched through to the output may still appear, in greatly attenuated form, at the output. The amount of *cross-talk* in a multiplexer can be determined by applying a full-scale input signal to an off-state switch, and measuring the fraction of this signal appearing at the output. Some manufacturers will express this fraction as a percentage, others as a ratio in decibels.

A complete analogue-to-digital interface system

In this chapter we have identified the main constituents of an analogue-to-digital interface, now we want briefly to examine a complete subsystem. Figure 96 is a block diagram of a typical interface subsystem. Notice that as well as the multiplexer, sample-and-hold and A/D converter, we have also shown a signal conditioning/processing block. This latter block, which might include amplifiers, anti-aliasing filters, or both, is included within the subsystem to emphasize the fact that its operating parameters must be matched to the rest of the subsystem.

The final block in the diagram is the system sequencer, which controls the input channel selection and provides all the control signals for the other blocks. Depending upon the application, this might be a general-purpose computer, or, a purpose-built unit. The two most important parameters for the total system are accuracy and conversion rate. Let us examine them individually, but always remember that they are inter-related, because higher conversion rates invariably mean reduced accuracy and high accuracy requires long conversion times.

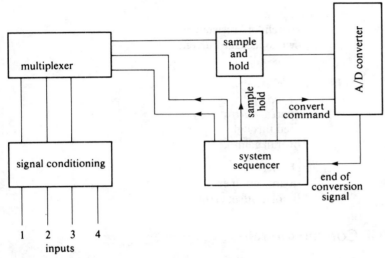

Figure 96   Block diagram of a multichannel interface subsystem

System accuracy

The first problem that is usually encountered in an analogue-to-digital conversion subsystem, is that the full-scale voltage of the input signal does not equal the full-scale input voltage of the subsystem. For example, suppose a temperature transducer gives a full-scale output of 2 V, and the subsystem will accept a 10 V full scale. If we were to connect the transducer directly to the subsystem, we would degrade the conversion accuracy because the quantization noise is proportionately larger. An example should make this clear. Assume that a 10-bit converter is to be used, then the quantization error is 1 part in 1024 for a full-scale input. If, however, the maximum input signal is only 2 V, and not 10 V, only 1/5th of the possible A/D converter codewords will be used and so the quantization error will be 1 part in $1024/5 \simeq 205$.

To minimize the quantization error it is essential that small input signals are amplified so that their full-scale range matches that of the conversion subsystem. Large signals on the other hand must be attenuated to prevent damage to circuit components. Assuming that all the system elements are matched for full-scale input voltage, the main factors that determine the subsystem's accuracy are as follows:

(i)   *Signal conditioner*
        amplifier frequency response
        filter frequency response
        overall gain

(ii) *Multiplexer*
        switch 'on' resistance
        switch leakage current
        cross-talk

(iii) *Sample-and-hold*
        droop rate
        aperture error
        feed through
        overall gain

(iv) *A/D converter*
        quantization error
        bipolar offset error.

Conversion rates

The second subsystem parameter we mentioned was conversion rate (or 'system throughput' in technical jargon). In the simplest of systems we might assume that all events occur sequentially, in which case a complete conversion sequence would be:

1.   Select multiplexer channel and allow output to settle, then;

2.   Issue sample command to sample-and-hold and allow capacitor to charge;

3.   Issue hold command to sample-and-hold;

4.   Start A/D converter and wait for end of conversion.

The total conversion time is simply the sum of the individual times. Although simple, this scheme is very inefficient; let us show you why.

The first point is that the multiplexer channel is selected at the beginning of each cycle and held until the next one. We can, if the sample-and-hold feed-through error is small, improve things by arranging to select the next channel once the sample-and-hold is switched to the hold mode. This enables the multiplexer to switch and settle during the A/D conversion time. If the multiplexer settles within the conversion time, then the end of conversion signal can be used to switch the sample-and-hold into the sample mode. The time required by the sequencer to read and transfer the A/D converter output data will now overlap the acquisition time of the sample-and-hold, and so may reduce the conversion time even further.

These two examples demonstrate how the throughput rate can be increased by overlapping the various events in the conversion chain. Again it is a question of using our knowledge of the subsystem blocks to achieve a specified throughput, rather than just employing the fastest available modules operating sequentially.

Summary

This chapter has described some of the devices and circuits used in the design of digital-to-analogue and analogue-to-digital conversion subsystems. The *digital-to-analogue (D/A) converter* is a device that converts a binary codeword to an analogue voltage or current. The output may be either *unipolar* or *bipolar*. A resistor *weighting network* determines the proportion of the output voltage controlled by each bit of the input codeword. The two most important weighting networks are the *binary weighted*, and the *R–2R ladder* network.

High performance D/A converters require a *stabilized voltage reference source* and switches to connect this source to the appropriate resistors in the weighting network. The introduction of the switches though produces the errors associated with *on-resistance* and *leakage current*. The conversion from digital to analogue or analogue to digital involves the *quantization* process which limits resolution and introduces quantization noise.

Analogue-to-digital (A/D) converters are used to obtain a binary representation of either voltage or current. The two types examined are the *counter-ramp* and *successive-approximation* converters. The conversion of d.c. signals is straightforward, but for a.c. signals we must *sample* the input. The rate at which we sample is determined by the *sampling rule* which requires that a signal of bandwidth f Hz be sampled at a rate in excess of $2f$ samples per second. If the A/D converter cannot maintain this rate of conversion it is essential that the input signal bandwidth be restricted, even though we may lose information, to prevent *aliasing* errors.

To prevent errors from variations in the input signal during the

A/D conversion cycle, a *sample-and-hold* is used to sample the signal and present a d.c. level to the A/D converter. The operational characteristics of a sample-and-hold are defined by the *aperture* and *acquisition* times, and the *droop-rate* and the *feed-through*.

The last device described in the chapter is the *multiplexer*, which enables one fast A/D converter sample-and-hold combination to be switched between several narrow bandwidth signal inputs. Provided that the sampling rate for all the input signals meets the requirements of the sampling rule, *time-division multiplexing* may be utilized to reduce overall system costs.

Problems for Chapter 4

1. An 8-bit D/A converter of the binary weighted resistor type shown in Figure 67 is to be designed using an electronic switch with an on resistance of 300 Ω. If the overall accuracy of the device is to be 0.5 per cent of full scale output, what is the minimum value of R that can be used?

2. What will be the output of the D/A converter shown in Figure 72 for input codeword 1010, if $R = 5$ kΩ, $-V_s = -5$ V and $R_f = 10$ kΩ?

3. What is the maximum quantization error for a 10-bit bipolar D/A converter with a full-scale output ± 10 V?

4. An 8-bit counter ramp A/D converter utilizes a clock with a period of 10 μs. What is its maximum conversion rate for a full-scale input signal?

5. What increase in speed can be gained by using a 12-bit successive-approximation converter as opposed to a 12-bit counter-ramp design assuming a full-scale input voltage?

6. What is the maximum bandwidth of signal that can be converted by an A/D converter with a converstion time of 0.25 ms?

7. What is the maximum size of capacitor that can be used in a sample-and-hold, if the switch resistance is 300 Ω and the maximum settling time to 0.01 per cent is 10 μs?

CHAPTER 5

Digital Components and Systems

Introduction

At the time of writing this book, a significant change in occurring in the way complex electronic systems are being designed and implemented. This statement applies in particular to digital systems, and the change is being brought about by a component called a microprocessor. The aim of this chapter is to take you towards an understanding of this change. Once you have read it you will have some idea of what a microprocessor is and will have a fuller grasp of what the terms ROM and RAM mean and of how components are interconnected by *buses* (and why). You should be aware of the difference between a microprocessor and a microcomputer, between an address bus and a data bus, between a volatile and a non-volatile memory. You will certainly *not* be equipped to design a microprocessor-based electronic system, nor will you be equipped with the engineering skills necessary to turn a functional system diagram into a workpiece of hardware. To obtain those skills requires further studies, but this chapter will prepare you well.

A microcomputer data collection system

To give a context to the whole discussion, we will consider the application of a microcomputer to an industrial data collection system. A manufacturer of car batteries has a quality control section which tests 100 batteries at a time by discharging them through load resistors. A typical voltage/time curve for a battery is shown in Figure 97. The battery voltage remains almost constant for about 4 hours, then rapidly falls off when the battery is completely discharged. The information required on each battery is the length of time for which the full voltage has been maintained, and detailed information about the shape of the failure characteristic. Each battery voltage is monitored using the system shown in Figure 98. A 100-way multiplexer selects each battery once every second and an A/D converter generates a digital version of the battery voltage. A system controller controls the multiplexer and A/D converter operation, and reads the digital output of the A/D converter.

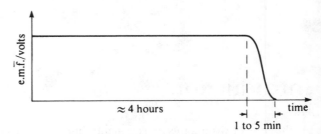

Figure 97

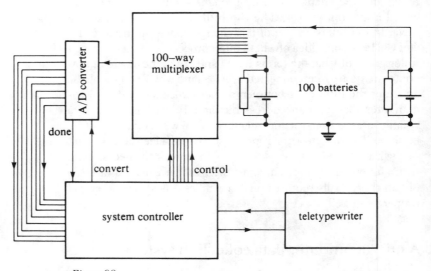

Figure 98

The battery voltage has to be read every second so as to obtain detailed information about the failure mode when it finally occurs, but all the readings taken cannot be recorded on the printer. If

122

they were the amount of data would be immense, and analysis of it would be practically impossible. For 100 batteries sampled each second for four hours the number of readings taken would be 100 × 4 × 60 × 60 = 1 440 000! Instead, the controller is required to print out data only when the reading of a particular battery voltage has changed by at least 50 mV from its previous read value. When the controller detects such a change, it must initiate the printing of the multiplexer channel number, the time and the new voltage level. In this way the amount of printed data is reduce to manageable proportions. The teletypewriter must also provide the means by which a human operator can record the time at which each battery is put onto the test system.

It is worth noting a few more details about the battery test system. The multiplexer will need to be provided with 7 control bits so as to allow any one of 100 input lines to be selected ($2^7 = 128$). The A/D converter only needs to have a quantization interval of 50 mV in a full-scale reading of 12 V. This is one part in 240, so an 8-bit output word is adequate ($2^8 = 256$). The maximum rate of change of battery voltage at failure is about 12 V in 30 seconds, or $1/5$ V s^{-1}. Assuming an A/D conversion time of 1 ms, the maximum voltage change which can occur during conversion is 0.2 mV. Since the quantization interval is 50 mV, a sample-and-hold unit is not required. Because all the batteries have very nearly the same steady-state voltage output, signal conditioning is not required before the multiplexer. We are not going to concern ourselves with the detailed design of such a system, but merely with one way in which the system controller could be implemented using a microprocessor. However, before we get to that discussion, we need to look in some detail at the components which will be part of *any* microprocessor system. The next part of the chapter will be concerned with these components.

A simple memory

Introduction

The system controller for the battery test system is required to 'remember' a voltage reading on a battery until the next reading of that battery is taken so as to be able to compare the two values and decide whether a data print-out is required. We are going to start with a discussion of how such a memory facility can be provided.

Multiple data registers

In Chapter 3 you met the D-type flip-flop, and the n-bit register. The flip-flop was a simple memory element which, when combined with other flip-flops, created a register. Thus, a flip-flop is an element of a register. In these units, we want you to think of the register as an element of a memory; that is, many registers combined together to form a simple memory.

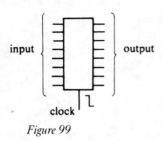

input   **output**

clock

Figure 99

The symbol we will use for such a 8-bit register is shown in Figure 99. We will assume that the flip-flops respond to the $1-0$ transition of the clock input as indicated on the figure; that is, at the next $1-0$ transition of the clock, the states of the input lines are stored in the flip-flops, and are available at the outputs. The states of the input lines can then change without affecting the outputs until the next $1-0$ transition of the clock. If we had four such registers, they would provide the facility for storing four 8-bit binary words. Now, suppose the binary words which are to be stored are all generated by the *same* 8-bit source; in other words, the input lines to the four registers are all connected in parallel, as in Figure 100. (Notice how we have tried to clarify the figure by drawing only two lines instead of eight; the two lines represent a pathway of eight individual wires linking common connections on each device. This is a frequently used convention. The arrow built into the pathway indicates the direction in which the digital words move from the source to the destination, and the number written in the pathway indicates the number of connections.) If we now also connected all the clock inputs together, then at a $1-0$ transition of the clock, the *same* digital word would be stored in all the registers. This is not very useful, if there are four registers, then they can be more usefully employed storing four *different* words. The clock pulse needs to be applied to one register only, even though the input word is applied to them all. Only the register receiving a $1-0$ clock transition will then store the input word. The source must generate a second piece of digital data called an

address   *address* which must uniquely specify which register is to receive the current data word output. Each register must be provided with an

address decoder   *address decoder*, which will permit a $1-0$ transition to occur at the clock input of the register *only* when the address has the value corresponding to the address allocated to that particular register. By giving each register a unique address (and hence a unique address decoder), the address information generated by the source can ensure that only one register receives each 8-bit word.

With four registers to be uniquely addressed, how many bits will be required in the digital address word generated by the source of data? For four registers, the address word will need two bits. Two binary digits have four unique combinations (00, 01, 10 and 11).

124

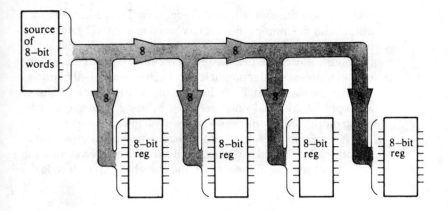

Figure 100

Register address decoding

We will now describe in detail how the addressing of the register can be achieved. At this stage we will not consider how the address information is generated by the source of data, but only how that information is decoded at each register.

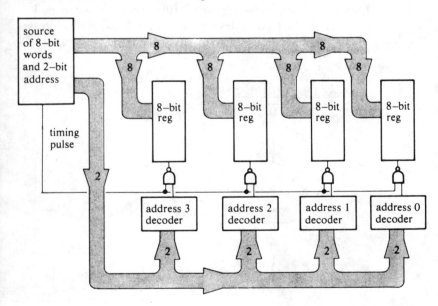

Figure 101

Look at Figure 101. It shows the interconnection of the source and the registers, including the connection of the address information. The address is generated by the source and is connected to an address decoder associated with each register. The source also generates a timing pulse which is used to synchronize the operations of the source and of the register. The address decoder output and the timing pulse are combined in a NAND gate, whose output can only become logic 0 when *both* inputs are logic 1. When the timing pulse goes to logic 1, only the clock input of the register whose address decoder output is also 1 can receive a 1–0 transition to enter the data word. Each decoder could be implemented using a simple ROM (read-only memory), by using logic gates, or by using more complex devices which we will introduce later.

The function of each decoder is most conveniently expressed as a truth table. Consider the action of the address decoder on the right-most register of Figure 101, the one which has to decode the address 00.

Its truth table will be:

Address MSB LSB		Decoder output
0	0	1
0	1	0
1	0	0
1	1	0.

Only when the address bits are 00 will the decoder output be 1.

Similarly the truth tables for the address decoder for addresses 01, 10 and 11 are:

Address 1	MSB LSB		Decoder output
	0	0	0
	0	1	1
	1	0	0
	1	1	0

Address 2	MSB LSB		Decoder output
	0	0	0
	0	1	0
	1	0	1
	1	1	0

Address 3	MSB LSB		Decoder output
	0	0	0
	0	1	0
	1	0	0
	1	1	1.

Each of the truth tables (and hence each of the decoders) can be implemented using a two-input, one-output ROM having the required transfer function. Alternatively, the truth tables can be implemented using a single two-input, four-output ROM. Another way of implementing the truth tables is to use logic gates. You met all possible combinations of a two-input gate in Chapter 2, and one or other of those combinations is identical with each of the decoder truth tables. For the address decoder, the truth table represents the AND-gate, as shown in Figure 102(a). (If either input is logic 0, the the output is logic 0. Only if both inputs are logic 1 is the output logic 1.) For example, the address 0 decoder is simply the NOR function of the two inputs, and could therefore be implemented using a two-input NOR gate as in Figure 102(b). In practice, because gates are usually supplied in packages containing several gates of the same type, it is common to use as few different types of gate as possible. Although the decoders for the registers can be made from three two-input AND gates, two inverters and one NOR gate, it would be normal to implement all four using only AND gates and inverters as shown in Figure 103.

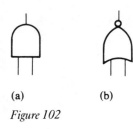

(a)   (b)

Figure 102

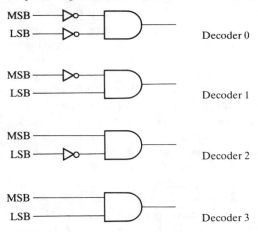

Figure 103

That, for the moment, completes our consideration of the address decoding. The complete decoding can be implemented using four ROMs each having two inputs and one output, or by using 2-input NAND gates and inverters. Equally of course a single ROM having two inputs and four outputs could achieve the decoding. Obviously, if there are more than four registers more address bits will be needed, and the address decoders will become more complex. However, no new principles will be involved.

This is an appropriate point at which to introduce another technical term used in connection with such a memory, the word *bus*. A bus in this context is a collection of wires or connections which serve more than one device. For example in Figure 101, the

127

eight lines which connect the source of data words to all the registers would be called an *8-bit data bus;* 8-bit because it actually contains eight individual lines each carrying one bit, and data because that is the information which passes along it. Similarly, in the same figure, the address information is carried by a *2-bit address bus.*

Selecting and combining register outputs

We want now to look at the output end of our simple memory. We are going to forget about the input and the need for address decoding just discussed, and consider what must happen if the *outputs* of the four registers have to be connected together so that any one of the register contents can be read by a single receiving device. This time the receiving device must be able to specify which register the data is to be read from. There will be a need for the receiver to generate a 2-bit address, and for each register to be equipped with an address decoder. But what must the output of the address decoder do? It must be used to put the data contained in that one register onto the eight wires leading to the receiver input data connections. Figure 104 shows what is required. The eight output lines of each register must pass through some form of 8-line switch, so that the data contents can be connected to or disconnected from the receiver depending on the output from the address decoder allocated to that register. Because of the need for fast, reliable operation of digital devices, these switches cannot be mechanical or electro-mechanical. They must be electronic, so that they can function at the same speed as other digital devices.

tri-state devices   One method of implementing these switches is to use *tri-state devices.* A tri-state device is one in which the output of the device can be in one of *three* states. Two of the states are identical to the two states of the output of a normal logic gate, that is the output is either at a voltage level representing logic 1 or at a voltage level representing logic 0. In the third state, the device is effectively disconnected from the bus, its output being it its *high-impedance state.* As a visualization of the three states, the internal circuit driving the output can be considered to be two transistors in a 'totem-pole' arrangement as in Figure 105. (The arrangement is called totem-pole because one transistor 'sits' on top of the other, just like human figures on an Amerindian totem pole.) When the output is in the logic 0 state, the lower transistor is switched on and conducting whilst the upper transistor is switched off. The output voltage is then the saturation collector-emitter voltage of the lower transistor (about 0.4 V). When the output is in the logic 1 state, the upper transistor is switched on and the lower transistor is switched off. The output is connected to +5 V through the upper transistor and resistor, and sits at about +3.6 V. In the high-impedance state *both* transistors are switched off, and the output voltage depends on external circuit components, not on the input to the device. To implement the switches on the register outputs *tri-state buffers* would normally be used. A tri-state buffer has two inputs, an 'output enable' input which switches off and on the high impedance state, and a logic input which determines the logic output state of the device when not in the high impedance state. Thus when the buffer is *enabled* the logic level of its input appears at its output. When it is *disabled*, the output is in its high impedance state. It is called a buffer because its output is capable of supplying the current requirements of the inputs of many other logic devices.

tri-state buffer

Figure 104

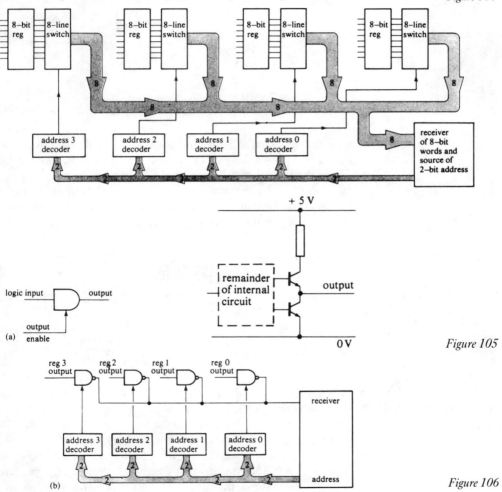

Figure 105

Figure 106

The symbol for a non-inverting tri-state buffer (that is, one in which the logic state of the output is the same as the logic state of the input) is shown in Figure 106a. Figure 106b shows the connec-

129

tion for one bit of each of the four registers. With this connection of tri-state buffers, only *one* of the tri-state devices whose outputs are joined together must be enabled at any time. If more than one has its output enabled, then spurious data can be sent to the receiver. The problem arises if two buffers are enabled at the same time, and their outputs are not the same. Figure 107 shows two totem-pole arrangements connected together with one in the 1 state and the other in the 0 state. Now the +5 V is connected to the 0 V line

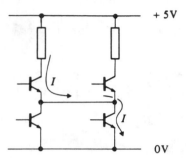

+ 5 V

0 V

Figure 107

through the two transistors and one resistor. In this condition the output voltage of the pair of buffers cannot be guaranteed to be either the logic 0 or the logic 1 level. Using the address decoders to enable the buffer outputs ensures that only one set of eight buffers is enabled at any time, and hence that the receiver gets the contents of only the selected 8-bit register.

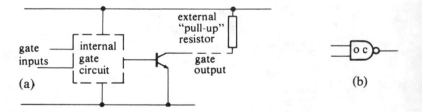

Figure 108

open-collector NAND buffer

An alternative device which can be used to implement the 8-way switches is the *open-collector NAND buffer*. Open-collector means that the internal fabrication of the gate is such that its output circuit looks like a transistor, having its emitter connected to the 0 V supply, and having its collector open circuit (Figure 108a.) The NAND means that when *both* inputs to the gate are logic 1, the base of the output transistor is driven to a positive voltage, turning the output transistor on, and connecting the output to the 0 V supply (that is, producing a logic 0 output in the convention usually used for gates. When *either* of the inputs is at

130

logic 0, the base of the transistor is driven to 0 volts, turning the transistor off and allowing the output voltage to be pulled-up to +5 V by the external resistor. By the normal so-called positive logic convention the *on* state of the output transistor gives rise to a logic 0 output of the gate, while the *off* state of the transistor gives the logic 1 output of the gate. The symbol for such a gate is shown in Figure 108(b). Such buffers can have their outputs connected together, so that the outputs of several registers can be joined together to feed the receiver. Figure 109(a) shows two sets of buffers having their corresponding outputs joined, however, the principle can be extended to many such registers. Figure 109(b) shows the electrical circuit which is created for each bit of the 8-bit data word for the four registers of Figure 103. (Notice that

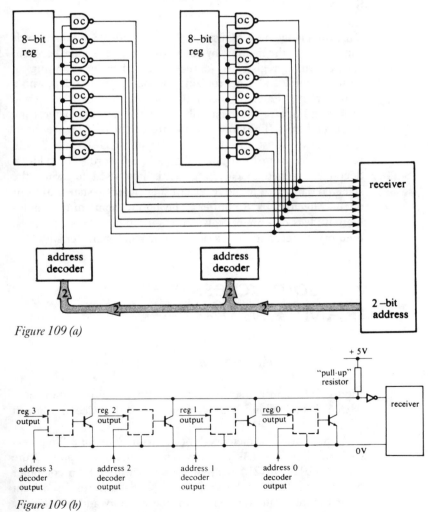

Figure 109 (a)

Figure 109 (b)

131

an inverter has been included at the input to the reciever.) Since only the address decoder output can be at logic 1 at any time (the address information can only be in one of its four possible states at any time) only one of the open-collector transistors can be *on*, and then only if the selected register content is also a logic 1. Therefore, if the voltage on the common line has been pulled down to 0 V a transistor must be switched on and so the data bit in the *selected* register must be a logic 1. The inverter changes this low voltage into a voltage of about $+3.5$ volts, which will represent a logic 1 input to the receiver. Thus, only the content of the register selected by the address information is passed to the reciever.

Summary

This then is the basis of a simple 8-bit memory. 8-bit data words can be sent to the memory by a source of data, the source having to specify which register is to receive the data by supplying an address word, and synchronizing the storage of the data with a timing pulse. Each register must have an address decoder to permit storage of the data only when the decoder detects the correct address. The stored data words can be read by a receiver. The receiver must also be able to specify which register it is to receive data from by means of an address word. The address decoder attached to each register is now used to switch the selected data word to the receiver using either tri-state buffers or open-collector NAND buffers. Having met some of the fundamental ideas of a memory, we now want to introduce a practical memory device, and to describe its functional characteristics.

Random-access memory (RAM)

A typical random-access memory device

Figure 110 is an extract from a description of a typical memory device. What follows is a description of this device and how it operates.

A random-access memory device contains 2048 (or 2^{11}) registers each of eight bits. (As you may remember from Chapter 2 the number 1024 (or 2^{10}) is often abbreviated to 1 K in computer parlance, so 2048 can be written as 2 K.) 8-bit words can be sent to or 'written' into any register of the memory and be taken or 'read' from any register. The term RAM (standing for Random-Access Memory) is only used to describe memory elements which can be *written to as well as read from*. It is called random access, because any of the 2K registers can be accessed for reading or writing by simply changing the address supplied to the device, and the order in which they are accessed is not fixed in any way by the device itself. Before starting to describe the device in greater detail, we want to make it clear that, as usual, we will *not* be concerned with the internal circuit configuration, but with the characteristics of the device as seen from outside.

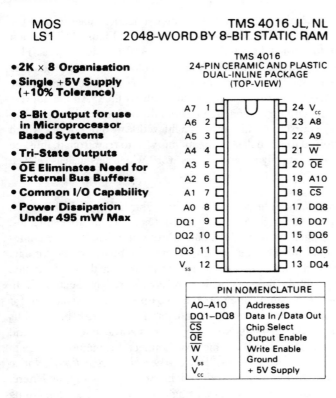

MOS
LS1

TMS 4016 JL, NL
2048-WORD BY 8-BIT STATIC RAM

- **2K × 8 Organisation**
- **Single +5V Supply (+10% Tolerance)**
- **8-Bit Output for use in Microprocessor Based Systems**
- **Tri-State Outputs**
- **$\overline{OE}$ Eliminates Need for External Bus Buffers**
- **Common I/O Capability**
- **Power Dissipation Under 495 mW Max**

TMS 4016
24-PIN CERAMIC AND PLASTIC
DUAL-INLINE PACKAGE
(TOP-VIEW)

A7	1	24 V_{cc}
A6	2	23 A8
A5	3	22 A9
A4	4	21 $\overline{W}$
A3	5	20 $\overline{OE}$
A2	6	19 A10
A1	7	18 $\overline{CS}$
A0	8	17 DQ8
DQ1	9	16 DQ7
DQ2	10	15 DQ6
DQ3	11	14 DQ5
V_{ss}	12	13 DQ4

PIN NOMENCLATURE	
A0–A10	Addresses
DQ1–DQ8	Data In / Data Out
$\overline{CS}$	Chip Select
$\overline{OE}$	Output Enable
$\overline{W}$	Write Enable
V_{ss}	Ground
V_{cc}	+ 5V Supply

Figure 110

Mechanical and electrical characteristics

The device outline and connections are shown on the figure. It is normally plugged into a corresponding 24-pin socket soldered to a printed circuit board. The package is called a dual in-line (or DIL) pack, which means simply that all the connections are arranged in two parallel straight lines. It requires a single +5 V supply, the +5 V being connected to V_{cc} (pin 24) and earth being connected to V_{ss} (pin 12). The only other electrical characteristic which is specified is the power dissipation, in this case it is guaranteed to be less that

495 mW, so that the current taken from the 5 V supply will be less than 99 mA. Power dissipation is important from two aspects, the size of the power supply required for the complete system, and the amount of heat which will have to be removed from it if individual devices are not to overheat. There are other important electrical characteristics, but we have chosen to omit them at this stage for the sake of clarity.

Functional characteristics

Device addressing: If you look at the connection diagram and the table below it labelled 'pin nomenclature' in Figure 110, the first line of the table tells you that pins A0 to A10 are address pins. This gives 11 address lines which are capable of 2^{11} unique address codes. Now $2^{11} = 2048$ so each individual 8-bit register can be specified by a unique code on the address lines. All the address decoding circuitry is contained within the device, we do not have to concern ourselves with how it is achieved. Any other device which is required to write data to the memory, or to read from it, must be capable of supplying an address word which has at least eleven bits.

The data lines: The second line of the table in Figure 110 says that pins DQ1 to DQ8 are data in/data out lines. There are eight lines, one for each bit of an 8-bit data word; but notice that these lines are used both for data being written to the memory, and for data being read from it. This is a major departure from the configuration of the simple memory which we considered in Section 2. In Section 2 the simple memory finished up with an 8-bit data bus for entering data words to the memory and a separate 8-bit data bus for taking data from it. Now only one bus is available for both, and this means the data bus which will connect this memory device to other devices in a system will have to be a *bi-directional bus*. (Notice how the term 'bus' introduced in Section 2 is applied here. Because the memory is now a single device, the bus is something which connects it to other devices, not something which connects its internal parts.) Because the data bus has to serve a dual function, a signal is reuired to control the device so that it performs only one function at any time. This signal is supplied to the memory on pin 21. It is called $\overline{W}$ (said 'W bar') and line 5 of the pin nomenclature table specifies the signal $\overline{W}$ as being a 'write enable' signal. (The words read and write can cause confusion unless you remember that they refer to the function of the memory as seen by devices outside it. Hence a write operation means that some external device is writing data into the memory, while a read operation means that some external device is reading data from the memory.) If the memory is required to accept a data word from the data bus, the signal $\overline{W}$ must be in one logic state; if the memory is required to put a data word onto the bus, the signal $\overline{W}$ must be in the other logic state. Since W stands for 'write', the bar over the letter indicates that the signal on pin 21 must be in its *low* state (that is, near to 0 V) if the memory is to accept data, and must be in its *high* state (about +3.5 V) if the memory is to send data.

In fact there are three signals which must be supplied to the memory to control fully its functions. These are $\overline{CS}$, the *chip select* signal, the $\overline{W}$ *write enable* signal and the $\overline{OE}$, *output enable* signal. The bar over each label indicates that the specified function is activated when each input is in the *low* state. These three control signals have a 'pecking order' of priority, meaning that one signal can override others. The chip select signal $\overline{CS}$ has highest priority, and unless the chip is selected, the other signals have no significance. The write enable signal $\overline{W}$ has second priority, and when write is enabled the signal $\overline{OE}$ has no effect. The memory has tri-state output buffers built into it and the signal $\overline{OE}$ switches those buffers into their high impedance state when it is *high*, and allows the contents of a memory register to be connected to the bus when the signal is *low*. The various functions and the required values of the control signals can best be summarized by a truth table. Instead of using 0's and 1's in the table, we will use H's and L's, H indicating that the control input is in a high logic state (about +3.5 V) and L indicating that the input is in a low logic state (near 0 V).

$\overline{CS}$	$\overline{W}$	$\overline{OE}$	Comment	Function performed
H	H	H		
H	H	L	Chip *not* selected,	
H	L	H		no function
H	L	L		
L	H	H	chip selected, write not enabled, output not enabled	
L	H	L	chip selected, write not enabled, output enabled	read
L	L	H	chip selected, write enabled	write
L	L	L		

As you can see from the truth table, there are only three distinct states of the device, no function, read and write, and so the same control could have been achieved with only two control signals (two signals having four unique combinations). Many memory devices only require two control inputs, this particular device is supplied with the additional control line because it has a spare pin available, and it is possible that its use can give rise to more convenient control in some system configurations.

This then is a typical memory component, and as you can see from one of the statements in Figure 110, the 8-bit format is particularly suited for use with a microprocessor. You will be meeting the microprocessor later on in this chapter, but for the moment it is worth noting that the vast majority of microprocessors at present in use (1982) use 8-bit words, and hence will interconnect conveniently with 8-bit memory elements.

Configuring memory devices to create large memories

Introduction

Digital systems using memory devices often have need for more memory registers than can be contained in a single device. For example, it is possible for an 8-bit microprocessor to generate a 16-bit address and hence to make use of 64 K (that is, 65,536) 8-bit memory registers. The largest RAM in production (1982) has 64 K bits. A very wide variety of RAM configurations is available, from very small (32 × 1 bit) to fairly large (8 K × 8 bit and 64 K × 1 bit). If a larger memory is required, then a means must be provided of uniquely specifying each memory address, and of combining the memory inputs and outputs so that they can all be connected to one bidirectional data bus. The way in which the devices are combined depends on the device configuration used, and on the size of memory which is finally required.

Using 2 K × 8 memory devices to create a 4 K × 8 memory

We have deliberately chosen a very simple example to start with so that the basic principles can be absorbed before some of the complexities. The combining together of two similar devices to produce a memory twice as big is made very easy by the *chip select* facility which is provided on each device. We will assume that we are using the memory device described by Figure 110 and that the source and receiver of the 8-bit words is capable of generating the required 12-bit address. Eleven of those bits are used to drive the address selectors contained within the devices, the twelfth bit is used to enable or disable each device as required using its chip select input $\overline{CS}$. Figure 111 shows the arrangement when there is only one device which is both the source and receiver of data words. Additional sources and receivers could be attached to the buses, but they would only complicate the basic discussion at this stage. Since one deivce is both the source and receiver, it must be able to indicate whether it is reading or writing by a signal **R/W** which we will assume is *high* for reading and *low* for writing. That signal is connected to the $\overline{W}$ input of both memory devices. The $\overline{OE}$ input of both memory devices is not required in this configuration, so we have shown it held at a low voltage to enable the tri-state outputs when $\overline{CS}$ and $\overline{W}$ are in their appropriate states.

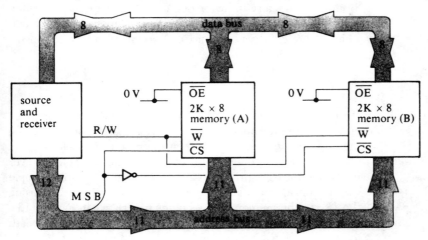

Figure 111

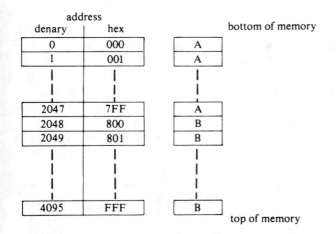

Figure 112

When the most significant bit of the address is 1 (the *high* logic state for address lines), memory A will be disabled by the high state on its $\overline{CS}$ input. This same address bit is inverted and fed to memory B, so that it *will* be selected when the address bit is 1. On the other hand when the most significant address bit is 0, it is memory A which is selected and memory B which is disabled. So

137

all addresses from 0 to 2047 refer to registers in memory A, while addresses 2048 to 4095 refer to memory B. This configuration of the memory is often shown diagrammatically by means of a *memory map*. Figure 112 is the memory map for the arrangement of Figure 111. It shows that the lower half of the address space is allocated to memory A and the upper half to memory B. The addresses have been shown in both denary and hexadecimal form. Putting the 'bottom' of memory at the top of the memory map may be slightly confusing, but it is a commonly used convention, and is the one which we prefer for reasons which cannot conveniently be explained.

A more complex example

If the task of creating a larger memory becomes more complex, either because the final memory needs to be larger, or because the memory devices are smaller, then the problem of selecting the correct device becomes more complex. The solution however uses the same basic principles already introduced. In this example we will consider how an $8K \times 8$ bit memory can be created using the $1K \times 4$ bit memory devices of the type shown in Figure 113.

The first point to notice is that we will need two memory devices to create each 8-bit word, and that the combination of them will provide $1K \times 8$ bit words. To create the whole memory we will need 8 pairs, or 16 memory devices. Each pair of memories will have the configuration shown in Figure 114. Each memory of the pair requires the same 10-bit address. The 8-bit data bus will be divided so that each memory is connected to only four of the lines in the bus. Both memories will be selected or disabled together, and they will both be either read from or written to at the same time. To simplify later figures we will represent the combination of two devices by a single box requiring an 8-bit data bus, a 10-bit address bus and two control signals $\overline{CS}$ and $\overline{W}$. The devices have tri-state data outputs, even though it is not stated in Figure 113. The high impedance state is selected when $\overline{CS}$ is in its high state.

For the complete $8K \times 8$-bit memory, the source and/or receiver must provide a 13-bit address word ($2^{13} = 8192 = 8$ K). Ten of the bits will provide the address information required by each memory device, while the other three bits need to be decoded to select one of the eight pairs of memories. The address decoder will need to have three inputs and eight outputs, one output only being *asserted* for each of the eight possible input codes. (The word 'asserted' is used in this context to mean that an output is in its active state, that is selecting the memory to which it is connected. For the example we are considering, the CS input to a memory has to be *low* to select the device, so a decoder output is asserted when

Pin Configuration

A$_N$ Address Inputs
I/O$_N$ Data Input/Output
$\overline{CS}$ Chip Select
$\overline{WE}$ Write Enable
V$_{ss}$ Ground
V$_{cc}$ +5V Power Supply

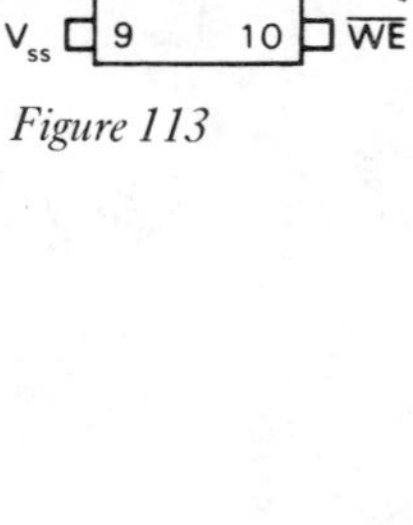

A$_6$	1		18	V$_{cc}$
A$_5$	2		17	A$_7$
A$_4$	3		16	A$_8$
A$_3$	4		15	A$_9$
A$_0$	5		14	I/O$_1$
A$_1$	6		13	I/O$_2$
A$_2$	7		12	I/O$_3$
$\overline{CS}$	8		11	I/O$_4$
V$_{ss}$	9		10	$\overline{WE}$

Figure 113

it is low. In another context, a signal might be in the high logic state for its line to be termed asserted.) Figure 115 shows the arrangement required, where each memory box represents the pair of memory devices shown in Figure 114.

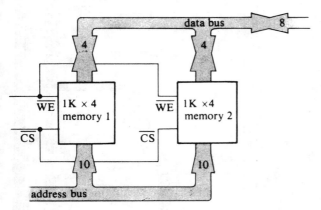

Figure 114

Address decoding

For the address bus lines, which provide the inputs to the decoder, we will assume that logic 0 is the *low* state of the line and that logic 1 is the *high* state. The outputs of the address decoder will have to enable just one 1 K × 8 memory for each combination of the three address line states. The truth table for the address decoder then becomes:

Address bits			Outputs							
MSB		LSB	0	1	2	3	4	5	6	7
L	L	L	L	H	H	H	H	H	H	H
L	L	H	H	L	H	H	H	H	H	H
L	H	L	H	H	L	H	H	H	H	H
L	H	H	H	H	H	L	H	H	H	H
H	L	L	H	H	H	H	L	H	H	H
H	L	H	H	H	H	H	H	L	H	H
H	H	L	H	H	H	H	H	H	L	H
H	H	H	H	H	H	H	H	H	H	L.

One method of implementing such a truth table is by means of a 3-input, 8-output ROM having the transfer function which matches the truth table. Indeed such a truth table is so commonly required that a special-purpose device, called a *3 to 8 decoder*, is manufactured to implement this truth table.

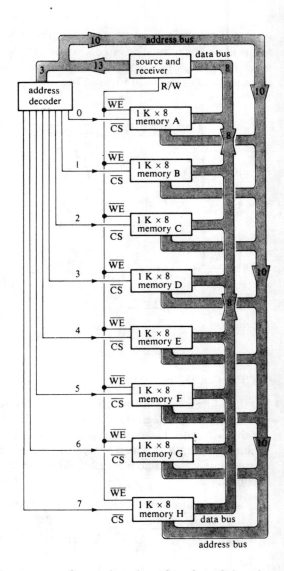

Figure 115

Figure 116 is an extract from a data sheet for a 3-to-8 decoder showing the pin connections of the device and its truth table. The inputs A, B and C are the inputs to which the 3-bit address will be supplied, A being the least significant bit and C being the most significant bit. Notice that the 3-to-8 decoder has three additional inputs G1, G2A and G2B. These are enable inputs for the device and their function is defined by the table in Figure 116.

This table is a concise version of a full 6-input truth table, and such a shortened version is often referred to as a *function table*. The first line of the function table says that if the G1 input is *low*, then whatever input is supplied to G2A and G2B, and whatever the address information, all the outputs will be in the *high* state. Similarly, the second and third lines tell us that if either G2A or G2B is *high*, then all the outputs must be *high*. Only when G1 is *high* and G2A is *low* and G2B is *low* can any output be *low*. Which output line is driven *low* is then determined by the address information. Just as we discovered earlier that memory devices can be enabled or disabled, so the 3-to-8 decoder can be enabled or disabled. By using more address bits to drive these enable inputs, several 3-to-8 decoders can be controlled so as to allow even larger memories to be created than the example we have just considered.

Figure 116

function table

inputs						outputs							
G1	G2A	G2B	C	B	A	Y0	Y1	Y2	Y3	Y4	Y5	Y6	Y7
L	X	X	X	X	X	H	H	H	H	H	H	H	H
X	H	X	X	X	X	H	H	H	H	H	H	H	H
X	X	H	X	X	X	H	H	H	H	H	H	H	H
H	L	L	L	L	L	L	H	H	H	H	H	H	H
H	L	L	L	L	H	H	L	H	H	H	H	H	H
H	L	L	L	H	L	H	H	L	H	H	H	H	H
H	L	L	L	H	H	H	H	H	L	H	H	H	H
H	L	L	H	L	L	H	H	H	H	L	H	H	H
H	L	L	H	L	H	H	H	H	H	H	L	H	H
H	L	L	H	H	L	H	H	H	H	H	H	L	H
H	L	L	H	H	H	H	H	H	H	H	H	H	L

H = high voltage level
L = low voltage level
X = don't care

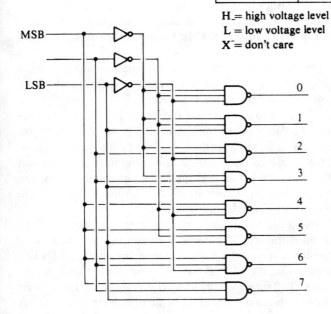

141

Timing considerations of RAM usage

We have seen how large memories can be created from smaller memory devices, and how the addressing of individual registers within the memory is achieved. All the discussion to date has been concerned with what might be called a 'combinational logic' implementation of the memory. We have been concerned with getting the correct logic levels on the correct lines so that writing to and reading from a memory can be accomplished. However, because they contain registers, memory devices are in fact sequential in nature. Things have to happen in the correct order if they are to function properly. We must therefore think about the relative timing of all the signals, as well as their logical correctness.

dynamic RAM

It is worth mentioning that there are two basic families of RAM devices, dynamic RAMs and static RAMs. A *dynamic RAM* is one in which the data bits put into the memory are stored as a charge on a very small capacitor. This charge leaks away, and so each memory cell must be 'refreshed' at regular intervals, typically once every few milliseconds. These 'refresh cycles' must be generated by additional circuitry external to the memory devices, and must be interleaved in time with reading and writing operations in the memory. Because such devices require quite complex additional circuits to enable them to function, we will not discuss them further in this book. However it is worth remembering that they exist, and are used extensively. A *static RAM* is one in which each data bit is stored as the state of a flip-flop, and that state remains indefinely, without refreshing, as long as the power supply to the memory device remains on. Such memory devices are very much easier to use than dynamic RAMs, but at the time of writing cost more.

static RAM

timing diagrams

Even static RAMs have critical timing requirements if they are to function correctly, and manufacturers normally specify these requirements by means of *timing diagrams*. To see exactly what a timing diagram consist of, we have reproduced exactly a manufacturer's timing diagram for a static RAM in Figure 117. The figure contains a timing diagram for the read operation, a timing diagram for the write operation, and a table of minimum or maximum values (which ever is the critical one) of the times indicated on the diagrams. At first sight they look very confusing, but they do become comprehensible with a little use.

Look first at the read cycle timing diagram. The top line of the diagram shows the instant A at which the address data reaches the state at which it specifies a particular memory register. This line also indicates that the address data must not change until instant B. The time $t_c(rd)$ for which the address data must remain constant is specified by the first entry in the table as at least 450 ns. (This part of the diagram therefore describes a requirement of the source of address data).

You will see the reason for this minimum time requirement on the address source very soon. The bottom line of the timing diagram shows that the data contained in the memory register will appear on the data terminals at instant E. The time which elapses between instant A and instant E, $t_a(A)$, has a maximum value (according to the fifth line of the table) of 450 ns. (This is a guarantee of the performance of the memory itself.) Now you should be able to see why the address data must be kept constant for at least 450 ns!

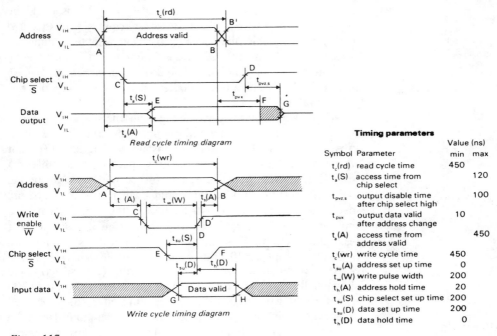

Read cycle timing diagram

Write cycle timing diagram

Timing parameters

Symbol	Parameter	Value (ns) min	max
t_c(rd)	read cycle time	450	
t_a(S)	access time from chip select		120
$t_{pvz.s}$	output disable time after chip select high		100
t_{pvx}	output data valid after address change	10	
t_a(A)	access time from address valid		450
t_c(wr)	write cycle time	450	
t_{su}(A)	address set up time	0	
t_w(W)	write pulse width	200	
t_h(A)	address hold time	20	
t_{su}(S)	chip select set up time	200	
t_{su}(D)	data set up time	200	
t_h(D)	data hold time	0	

Figure 117

The centre line of the timing diagram shows that the particular memory device in which the specified register exists should normally be enabled at instant C, *after* the address data has become valid. There is no minimum or maximum time specified for the time from A to C, so the diagram indicates that the constraint on the enable signal is that is should not occur *before* the address data is valid. The other constraint on the timing of the enable signal is shown by the time t_a(S) from instant C to instant E. This time is specified in the table as having a possible maximum time of 120 ns. Therefore, if the whole read cycle is to be achieved in a time

143

close to 450 ns, the chip enable must not be delayed after the address has become valid by more than 330 ns. To summarize the information extracted from the timing diagram so far, the address data should be supplied first, followed within 330 ns (if the read is not to be delayed) by the device enable signal. The register contents are then guaranteed to be stable on the data out terminals 450 ns after the address data has been supplied.

The other timings on the timing diagram indicate what can be expected to occur when the address changes again. Time t_{pvx} shows that the data out remains valid for at least 10 ns after the address change, and ($t_{pvz,s}$) that the data out will disappear (as the tri-state outputs return to their high-impedance state), within 100 ns of the removal of the enable signal. The 10 ns could be important if successive read operations are required to be executed as rapidly as possible, and hence t_c(rd) is being kept to its minimum possible value, and the 100 ns is important in indicating the minimum time for which the enable signal must remain *high* on this device before data from another selected device can be considered valid.

Now look at the write cycle timing diagram. Again it starts with the instant A at which the address becomes valid. The address must not *start* to change until the instant B. The time t_c(wr) for which the address must remain steady is specified in the table as at least 450 ns. The write enable signal (the second line of the timing diagram) must be supplied at instant C, which must not be before instant A, because t_{su}(A) is specified in the table as having a minimum value of zero. This write enable signal must last for at least 200 ns (t_w(W) = ns minimum) and the address data must be held valid for at least 20 ns after the write enable signal is removed (t_h(A) = 20 ns, minimum). The device enable (chip select) signal shown in line 3 of the timing diagram must be supplied at least 200 ns before the write enable starts to be removed (t_{su}(S) is specified as having a minimum value of 200 ns), and the input data to be written into memory (line 4 of the timing diagram must be present at least 200 ns before the end of write enable (t_{su}(D) = 200, ns minimum). Finally, the diagram shows that the data can be removed from the 'data in' terminals as soon as the write enable is removed (t_h(D) = 0 ns minimum).

These timing requirements seem to place severe constraints on the devices sending data to, or receiving data from, a memory device. However memory devices are generally used in conjunction with a microprocessor. The microprocessor acts as the source of data being written into the memory and as the receiver of data read from the memory. Such processors have been designed to satisfy the timing requirements of memory devices, and can frequently be connected to them without any thought of critical timings. The

144

time constraints are only of significance when a designer needs to check whether a given memory device is compatible with a given microprocessor and vice versa.

Summary

Random-access memory devices (RAMs) are memory devices containing many registers. The devices contain all address decoding necessary to address uniquely any individual register. The data connections to the memory device are used for both reading and writing operations in the internal registers and must therefore connect to a bi-directional data bus. The device requires a signal from the source and/or receiver of data to indicate whether a read or a write operation is required. All the data connections have internal tri-state (or sometimes open-collector) buffers to allow stored data to be connected to or disconnected from the data bus. An 'output enable' signal controls these tri-state outputs. The whole memory device can be enabled or disabled by a 'chip select' signal.

Memory devices can be interconnected to form larger memories. When they are so interconnected, a means must be provided of selecting the required device using some of the address bits generated by the source/receiver of data. ROMs, simple gates and special purpose decoders can be used to perform this selection. The particular use made of address bits in the selection circuit determines the 'position' of each memory device in the overall address space. A memory map is a diagrammatic representation of the allocation of devices to specific address areas. When very large memories are required, decoders such as the 3-to-8 decoder can themselves be enabled or disabled by 'chip select' inputs so as to allow the use of multiple decoders. Additional address bits are used to select a specific decoder, and hence to select the memory devices controlled by that decoder.

Memory devices place quite critical timing constraints on the source/receiver of data if the combination is to function correctly. Normally memories are designed to work with a specific microprocessor as the source and receiver of data. When using devices which are specified as compatible, the critical timing constraints have all been satisfied by the designer of the devices, and the system designer need not be concerned with them.

Read-only memory (ROM)

Random-access memories have one serious drawback for some applications, they are *volatile*. This means that when the power supply to the memory is switched off, all the data contained in it is lost. In many memory applications, the stored data is required day after day, and to be preserved during power down of the system. In large computers using volatile memory, the data is normally stored on magnetic tapes or discs prior to switching off, and then re-stored in the memory after switch-on. However, in many of the applications of memory the provision of such non-volatile back-up storage devices is just not economically feasible. To overcome this problem, data which can be determined before the system is put into operation can be sorted in a read-only memory (ROM). You have already met ROMs in Chapter 2 as devices which can implement a truth table. The idea of each combination of the input variables specifying an address in the ROM was also introduced. In the context of these units, the ROM is an alternative data storage medium to a RAM, and its function is to hold data which is required to be used over and over again, and which can be determined prior to the actual use of the memory in a system. Note that a ROM is also random access in nature, even though the term RAM is reserved for memories which can be written to as well as read from.

In this section, we want to revise some of the descriptions given in Chapter 2, the different types of ROM, and to enlarge on those descriptions. Later on we will be discussing, in the context of the microprocessor-based test system controller, the uses of each of these types.

Mask-programmed ROM

This type of ROM requires the data to be stored to be specified before the memory device is even manufactured. During the manufacturing process, a slice of silicon is subjected to a variety of chemical processes to create the required functional characteristics in the silicon slice. The chemical processing of the slice must be restricted to certain areas of it, different areas of the slice requiring different processing. The parts of the slice subjected to any specific process are determined by shining light through a photographic mask onto the slice surface. There, the light chemically alters a protective coating on the slice, so permitting or inhibiting reaction between areas of the slice and other chemical elements in its environment. One of those processes is the deposition of electrically conducting metal on to the surface of the slice to create an interconnection pattern between the various slice areas. This deposition of conductors is used to establish the data contents of a ROM, and since the deposition pattern is controlled by a photographic mask, such a device is called a *mask programmed ROM*.

What are the implications for the user of the device of this method of production? The first obvious implication is that the user must know what data is required in the memory a long time before he is anticipating using it in a system. Although the actual production of the silicon slices is a very rapid process, the production of the masks before production takes a considerable time, and the mounting of the silicon chips obtained from the slices into a robust package is also time consuming. As well as this, each specified memory must wait for its turn in a very busy production process. It is normal (in 1982) to have to wait months between placing an order for a mask-programmed ROM and receiving delivery of the completed order.

There is also a significant economic implication. The production of the masks is an expensive process, costing several thousand pounds. All this investment must be recouped by the manufacturer from the one customer who has placed the order. The cost per unit of the actual production and packaging is small, so the final cost to the customer of each ROM is determined very much by the quantity of identical units which are ordered. For mask programming to be economic, an order must be for thousands of identical units, rather than for hundreds. For large-quantity production however, mask-programmed ROMs are far cheaper than any of the alternatives.

One other significant implication for the user is that the specification of the data to the manufacturer *must* be correct (and it is the user who supplies this information). If there is an error in the specified data then the user will receive a large quantity of unusable memories for which he has to pay. The manufacturer normally supplies a few devices in advance to check that his mask-production techniques have not introduced errors, but if the stored data matches the specified data, the full mask costs must be paid by the user, even if the devices are not what he wanted.

Field-programmable read-only memory (FPROM)

A *field-programmable read-only memory or FPROM* is a read-only memory which is produced by the manufacturer without its data contents. The user can store his required data in the memory by applying electrical signals to the pins of the device which fuse internal connections (or leave them intact) to create patterns of 0s and 1s. This is a once-and-for-all process, and the memory contents, once programmed, cannot be altered. Because of the fusing of conducting links on the silicon chip, the devices are

fusible PROM

PROM programmer

sometimes called *fusible PROMS* (which also suits the abbreviation FPROM). The programming of the memory device requires the use of a *PROM programmer,* which is a special-purpose piece of equipment capable of receiving the required data patterns from for example, a minicomputer, and then rapidly applying the signals to the memory device to establish the required data contents.

For the user, the advantage of FPROM is that he can obtain, for the cost of one component, a ROM containing the required data for the system. If he discovers a mistake in the specification of the data, the only loss is the one FPROM device, and correction of the error involves only the programming of a second device. If only a relatively small number of systems is to be produced, the FPROM can be used to provide the ROM requirements of the production systems, thus avoiding the large overhead costs associated with mask-programmable ROMs. The PROM programmer itself represents a significant overhead cost, and for the small producer could be an uneconomic investment. To overcome this drawback, distributors of FPROM devices usually offer a programming service. For an additional charge they will use their own PROM programmer to establish the required data in one or more FPROM devices.

Erasable-programmable read-only memory (EPROM)

erasable-program-mable read-only memory EPROM

An *erasable-programmable read-only memory (EPROM)* is similar to an FPROM in that it is manufactured without data contents, and must be programmed by the user via a PROM programmer. However, the basic difference between the two is that the EPROM programming is not a once-and-for-all operation. If the data contents need to be changed they can be erased in preparation for re-programming. The erasure is achieved by illuminating the chip (through a transparent window in the package) with intense ultra-violet light from a suitable source. The ultra-violet light imparts sufficient energy to the silicon chip to disrupt electrical charge patterns implanted in it during programming, so destroying the stored data. For the user, the advantage of EPROMs over FPROMs is that a device does not have to be thrown away every time an error in data is discovered. The same device can be re-programmed to remove the error.

The disadvantages are firstly that EPROMs are generally more expensive than FPROMs, and secondly that they are not absolutely permanent stores of the data. They do rely on implanted charge in the silicon, and such implanted charges do leak away eventually, even though it may take years. The EPROM is normally used as a development aid to a finished product, the final product using either ROM or FPROM. Again the ultra-violet light source incurs a capital cost (not as great as that of a PROM programmer), and again distributors of the devices offer an erasing and re-programming service.

Summary

Read-only memory is used because it is non-volatile, that is it retains the stored data while the power is switched off. ROM is available as mask-programmed ROM, where the data contents are built into the device at the time of manufacture; as FPROM, where the user can insert his own data once and once only; and as EPROM, where the user can both insert his data and erase it for re-use if necessary.

All types of ROM are available in a variety of configurations, just like RAM. Typically $1 \text{ K} \times 8$, $2 \text{ K} \times 8$, $4 \text{ K} \times 8$, $2 \text{ K} \times 4$, $8 \text{ K} \times 4$, and so on, configurations can be obtained. ROMs are also supplied with tri-state outputs built in so that they may be individually enabled or disabled and so connected to a data bus to create larger memories. They can also be combined with RAM in a system to provide partly non-volatile memory and partly memory which can be written into.

Introducing the microprocessor

So far, in discussing memory devices, we have once or twice mentioned the word 'microprocessor'. In this section we are going to take a look at what a microprocessor is, and at some of its characteristics. We will not attempt to describe its internal structure either electrically or functionally. However, it is important for you to get some sort of feel for what this all-pervading device is, and how it links with other components to produce a micro-computer. A microprocessor is a very versatile receiver and source of data words read from and written into memory. Its versatility arises from the variety of ways in which it can use the data which it receives. Even with a wide variety of functions which it can perform, it is still an inherently simple device, in that its action is entirely determined by the data which it receives. Highly sophisticated functions which are attributed to microprocessors by the popular news media should in fact be attributed to the human who designs the data to be fed to the microprocessor. Because microprocessors vary so much from type to type, and because there are currently a large number of different microprocessors, we shall base this discussion on one processor which is one of the most commonly used devices. It is the Motorola MC6800 microprocessor.

Input–output signals

Figure 118 is reproduced from the manufacturer's literature for the MC6800. It shows all the inputs to and outputs from the device. Starting at the top right-hand side of the diagram you can see that the device has connections for an 8-bit bidirectional data bus, and so is capable of sending or receiving 8-bit data words. It also has connections for a 16-bit unidirectional address bus, from the microprocessor to other devices. This indicates that it is capable of generating 16-bit address words. At the bottom of the right-hand side is the read/write line, which was a signal we said earlier would need to be generated by a device which was both a source and receiver of data words. The remainder of the inputs and outputs (with the exception of the power supply connections at the top) are new to you, so we will briefly describe the function of each.

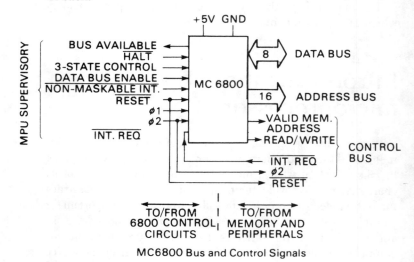

Figure 118

MC6800 Bus and Control Signals

VALID MEMORY ADDRESS is. a signal generated by the device to indicate to other devices to the address and data buses that address information has been issued by the MC6800, and that the microprocessor is about to receive or send data. This signal is usually used to enable other devices attached to the buses, for example to enable a memory device so that it performs a read or write operation, to generate or store the data.

150

$\overline{INT.\ REQ}$ is an input to the microprocessor and is more difficult to explain in terms of its effect *inside* the microprocessor, but as far as other devices attached to the buses are concerned, it is a line which, when asserted to a *low* voltage (hence the bar over the top) indicates to the microprocessor that some external device requires some action from it. Its full title is $\overline{INTERRUPT\ REQUEST}$, which indicates that the external device is telling the microprocessor that it should suspend its current function and perform some other function appropriate to that device. This is further discussed later in the chapter.

The last two signals on the right-hand side, φ_2 and $\overline{RESET}$, do not originate in the microprocessor itself, as might appear at first sight, but are two of the signals shown on the left-hand side of Figure 118 as supplied to it from some other external device. All these left-hand side signals are called *MPU supervisory* signals. MPU is the manufacturer's abbreviation for 'microprocessing unit' and stands for the microprocessor itself, and the word supervisory implies that they perform some overall supervision of the operation of the device.

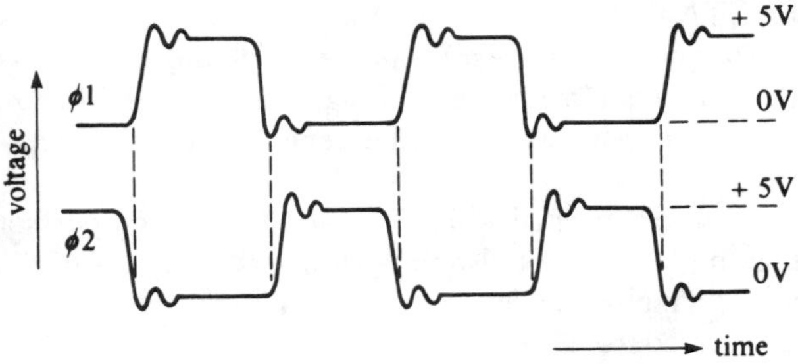

Figure 119

φ_1 and φ_2 are continuous square waves applied to the microprocessor to synchronize all the internal operations which it performs. The two square waves have to be at the same frequency, but antiphase to each other and non-overlapping (in that one falls to zero volts *before* the other rises from zero volts). Figure 119 shows typical waveforms for φ_1 and φ_2. The combination of the two form what is called a two-phase *system clock*. The waveforms are generated by a special device, or by a circuit made up of discrete components. For the MC6800, the clock frequency is specified as having to be between 0.1 MHz and 2 MHz. As is indicated on Figure 118 one of these square waves, φ_2, is also fed to other devices attached to the buses to synchronize their operations with those of the microprocessor. (Newer types of microprocessor have internal clock generating circuits.)

system clock

$\overline{RESET}$ is a signal which must be generated externally, and it causes the internal circuits of the microprocessor to perform a

151

predetermined initial function. It is required whenever the device is switched on, as otherwise it would perform in an unpredictable way. (We will return to this later in the chapter.) That signal too is fed to other devices on the buses to ensure that they are in a correct initial state after switching on the power. The signal is very often derived from the power switch-on circuit (for example by means of an *RC* network) to ensure that the $\overline{\text{RESET}}$ line remains low for a few milliseconds after the power has been switched on.

NON-MASKABLE INT. is another interrupt request line similar in function to $\overline{\text{INT. REQ}}$. The difference between the two need not concern us here, because we do not intend to discuss either at this stage.

DATA BUS ENABLE is a signal into the microprocessor which controls the internal tri-state outputs which put data onto the data bus. When 'data bus enable' is *low*, the tri-state outputs on the data bus will be in a high-impedance state. (The read/write signal generated within the microprocessor also controls those tri-state outputs. Before the microprocessor can put data on the data bus, the 'data bus enable' must be high, and 'read/write' must be in the 'write' state.)

THREE STATE CONTROL is an input signal which controls the tri-state outputs within the microprocessor driving the address bus lines and the read/write line. The signal must be *high* before address words or 'read/write' can be generated by the micro-processor.

HALT is an input signal which will stop the microprocessor performing any function. It also puts all tri-state outputs of the processor into a high-impedance state so that the device is effectively disconnected from the buses.

BUS AVAILABLE is a signal generated within the micro-processor. Under normal operating conditions, this line is kept in the *low* state. Whenever the $\overline{\text{HALT}}$ line is asserted, the 'bus available' signal goes *high*. The signal can be sensed by other devices to indicate that the microprocessor is not functioning.

Some basic functional characteristics

Perhaps the most important characteristic of a microprocessor is that it is a sequential device. It performs any required function by executing a *sequence* of steps which, taken together, produce the function. It is the system clock which determines the speed with which any sequence is performed by the microprocessor. Each individual step of a sequence involves the generation of an address, and the reading in of data from a memory device. Sometimes one step itself involves a short sequence of successive generations of addresses and the reading and/or writing of data from or to a

152

memory device. *The microprocessor itself cannot perform any function unless a memory device is also attached to its address and data buses.*

Any step in a sequence always begins with a memory read operation. The first data word which is read into the microprocessor is always used to control its subsequent operation. This is the characteristic which gives the microprocessor its inherent versatility. An 8-bit word can have 256 possible combinations of 1s and 0s which means that the first read operation could give rise to 256 separate subsequent actions. In practice, not all combinations are used, and the number of possible consequent actions is less that 256. However, some of the possible combinations will control the microprocessor to cause it to execute one or two more read operations, and to use the additional data words to control its operation. This obviously provides a very large number of possible actions. Some of those actions occur wholly within the microprocessor, other will cause read or write operations to other devices via the address and data buses. When we have looked at a microprocessor system configuration in the next section, we will illustrate a few of the steps which the microprocessor can execute.

A microprocessor system configuration

We are now in a position to consider how a microprocessor and memory devices can be interconnected to create the controller for the battery test system described in the introduction. The controller, when complete, must have the ability to control a multiplexed A/D converter and accept data from it; it must have the ability to store a voltage reading of each individual battery; it must have the ability to compare two readings and to have its subsequent action dependent on the result of the comparison, and it must be able to control the printing of information on a teletypewriter.

Earlier in the chapter we stated that *every* basic step which the microprocessor performs starts with a memory read to obtain an 8-bit control word, and hence a microprocessor cannot function without at least one memory device. Furthermore, since the first operation performed after switch-on of the processor is a memory read, the first word which is read must reside in a non-volatile memory device if the operation of the microprocessor is to be predictable. The system must therefore contain a ROM. Our application also requires the microprocessor to write A/D converter output data into memory for later use, so a RAM is also a requirement. The system will also require a power supply, a system clock and a start-up circuit to assert RESET when the power is first switched on.

There will also have to be some means by which the microprocessor communicates with the A/D converter and the multiplexer, and with the teletypewriter. Such a combination of components, together with the microprocessor itself, constitutes a *microcomputer*.

microcomputer

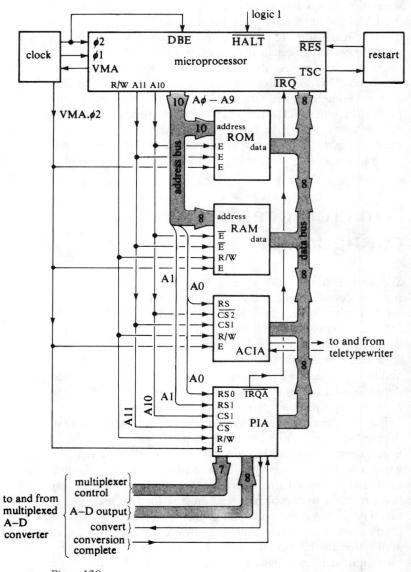

Figure 120

154

The specific configuration

The system components and interconnections

Figure 120 is a diagram of the system representing the interconnection of the devices and the signal paths in the system configuration. It looks very complicated at first sight (a failing of much technical literature on the subject of microprocessors), but you have met everything you need to understand it except for the boxes marked PIA and ACIA. The PIA (peripheral interface adaptor) is the device which allows communication with the multiplexed A/D converter, and the ACIA (asynchronous communications adaptor) allows communication with the teletypewriter. We will be discussing both of these devices in later sections.

Around the microprocessor there is a two-phase clock producing φ_1 and φ_2, and a restart device to hold $\overline{\text{RESET}}$ ($\overline{\text{RES}}$) low for a few milliseconds after switch-on. The same device holds the 'three state control' (**TSC**) input low for a short time after switch-on to prevent the generation of spurious addresses or read/write signals. The 'data bus enable' (**DBE**) signal is provided by φ_2 of the system clock so that the data bus can only have data put on it by the microprocessor when φ_2 is *high*. The system clock also accepts the 'valid memory address' (**VMA**) signal from the microprocessor and generates the signal **VMA**.φ_2 which is high only when both **VMA** and φ_2 are *high*. The signal **VMA**.φ_2 is used to enable all the devices attached to the address and data buses, thus synchronizing their operation with that of the microprocessor.

The system shows the data bus and the address bus connecting the microprocessor to the ROM, to the RAM, to the PIA and to the ACIA. The E inputs to the ROM, RAM and other devices are enable inputs, where a *high* state is required to enable the device. The E inputs to the RAM are also enable inputs, and are those for which a *low* state is required to enable the device. Any device is only enabled properly when *all* the enable inputs are in the correct *high* or *low* state. The ROM is enabled when VMA.φ_2, A10 and A11 are all in the *high* state, because each signal is connected to an E input of the ROM. The RAM is enabled when VMA.φ_2 is in its *high* state (it is connected to an E input), and A10 and A11 are in the *low* state (because they are connected to $\overline{\text{E}}$ inputs of the ROM). Address lines A15, A14, A13 and A12 are not used in this configuration, because the microprocessor can generate more addresses than are needed for this application. Notice that the ROM uses ten address lines so it must have a capacity of 1 K locations and the RAM with its eight address lines 256 locations.

The read/write signal from the microprocessor (R/W) is fed to the RAM, to the PIA and to the ACIA. The ROM does not need this signal because writing in the ROM is not possible. The interrupt request line ($\overline{\text{IRQ}}$) is not of interest at this stage of consideration of the system. Finally, the 'HALT' ('H') input to the microprocessor is held at logic 1 so that the microprocessor will function.

Just as we drew memory maps for large memory systems in an earlier chapter, so we can draw memory maps for complete systems like this one. The task becomes a little more complex with several devices, particularly when some of the address lines are unused. The total number of addresses which can be generated on a 16-bit address bus is 64 K (65 536). This is obviously far more than is required to address uniquely each location in the ROM and RAM in this case. The particular use made of the address lines in Figure 120 means that each location in the ROM and RAM can be accessed by more than one address. The memory map should indicate this, together with the addressing of the PIA and ACIA, the devices which implement communication with external devices.

The ACIA uses one address line (A0) together with the R/W signal to address four 8-bit registers contained inside it. Two of the registers can be written into by the microprocessor for transmission of data to the teletypewriter, and two can be read by the microprocessor for data in from the teletype, R/W selects read or write, and the state of A0 determines which register is selected. Address lines A11 and A10 connect to CS1 and $\overline{\text{CS2}}$ 'chip select' inputs which require one *high* state and one *low* state to enable the ACIA.

The PIA uses two address lines (A0 and A1) to select any one of the four 8-bit registers contained in it, the R/W signal (to indicate whether communication is out to the A/D converter or in from it) and two address lines (A10 and A11) to select (or enable) it using its CS1 and CS2 inputs. To create the address map, it is useful to identify clearly those address word bits which enable each device, those which address registers within it, and those which have no effect (and hence where the state of the line is irrelevant). This information can be represented in the table below.

Device	Address bits															
	A15	A14	A13	A12	A11	A10	A9	A8	A7	A6	A5	A4	A3	A2	A1	A0
RAM	X	X	X	X	0	0	X	X	A	A	A	A	A	A	A	A
PIA	X	X	X	X	0	1	X	X	X	X	X	X	X	X	A	A
ACIA	X	X	X	X	1	0	X	X	X	X	X	X	X	X	X	A
ROM	X	X	X	X	1	1	A	A	A	A	A	A	A	A	A	A

X = don't care (irrelevant)   A = address bit   1 or 0 = select bit

156

Check this table against the address line connections of Figure 120 until you are sure you understand the significance of each entry.

Because the values of A15, A14, A13 and A12 are irrelevant, the remaining bits of the address word will have the same effect in the system for all sixteen possible combinations of these bits. This means that the address map is repeated sixteen times in the 64 K of possible address space. We need therefore only consider the lower 4 K, and then repeat that allocation sixteen times. Looking at address bits A11 and A10, the first combination of them 00, covers 1 K of the 4 K memory area, all of which is associated with the RAM. 01 covers 1 K associated with the PIA, 10 covers 1 K associated with the ACIA and 11 covers 1 K associated with the ROM. The first breakdown of address space usage then looks like Figure 121. We now need to look at each of these 1 K areas in greater detail to establish the full memory map.

hex address	use
0000 to 03FF	RAM
0400 to 07FF	PIA
0800 to 0BFF	ACIA
0C00 to 0FFF	ROM

repeated 16 times to fill total address space of 64K

Figure 121

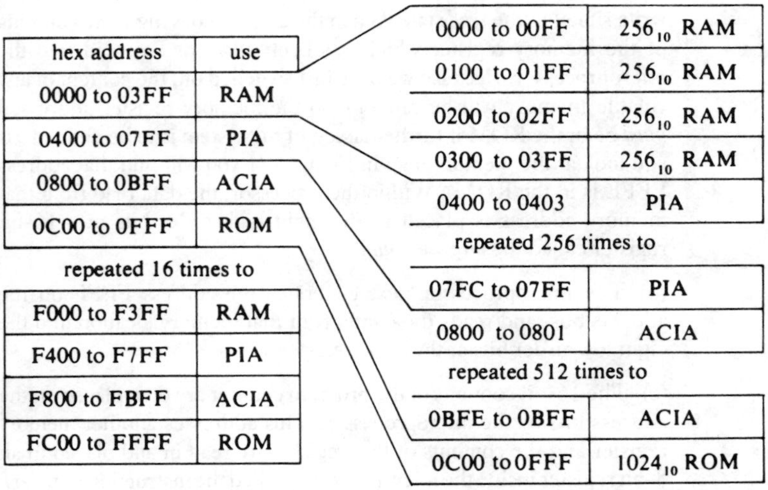

hex address	use
0000 to 03FF	RAM
0400 to 07FF	PIA
0800 to 0BFF	ACIA
0C00 to 0FFF	ROM

repeated 16 times to

F000 to F3FF	RAM
F400 to F7FF	PIA
F800 to FBFF	ACIA
FC00 to FFFF	ROM

0000 to 00FF	256_{10} RAM
0100 to 01FF	256_{10} RAM
0200 to 02FF	256_{10} RAM
0300 to 03FF	256_{10} RAM
0400 to 0403	PIA

repeated 256 times to

07FC to 07FF	PIA
0800 to 0801	ACIA

repeated 512 times to

0BFE to 0BFF	ACIA
0C00 to 0FFF	1024_{10} ROM

Figure 122

First the RAM area between hexadecimal addresses 0000 and 03FF inclusive. For this area, address bits A9 and A8 are not used, so their value is irrelevant. This means that for each one of the four possible combinations of these address bits, the remaining

157

bits (A7 to A0) will validly address the RAM register locations. As far as the map is concerned, the 256 RAM locations are repeated four times to make up 1 K area. Next the PIA area; this time address bits A9 to A2 inclusive are not used, so the address allocation established by bits A1 and A0 will be repeated 2^8 (256) times. A1 and A0 specify which of the PIA registers is being addressed.

In the ACIA area, address bits A9 to A1 inclusive are not used, so the two registers addresses selected by the address bit A0 will be repeated 2^9 (512) times. ROM addressing uses address bits A9 to A0 to specify any one of the 1024 ROM locations, which is the full 1 K of this first ROM address area. Figure 122 is the complete memory map for the system.

Some examples of specific microprocessor actions

The action of the microprocessor when first switched on

When the microcomputer is first switch on, the restart circuit will hold RESET *low* for a few milliseconds after the 5 V supply is fed to the system components. The effect on the microprocessor of this assertion of the RESET line is as follows:

(a) The microprocessor generates the address FFFE (hex) on the address bus, generates 'valid memory address', and puts the read/write signal in the *read* state. When the φ clock goes high, the contents of the memory register which has been addressed is read into the microprocessor. Because we have just switched on, the content of any volatile memory will be random, so the memory register addressed *must* be in the ROM if further action of the system is to be defined. (If you look at the memory map in Figure 121 you will find that address FFFE is in the ROM). Within the processor, the data read from that memory address is placed in the eight high-order bits of a 16-bit register called the *program counter*.

program counter

(b) The microprocessor next generates the address FFFF on the address bus, and reads the contents of that memory location into the eight low-order bits of the same register.

(c) The 16-bit contents of the program counter are then placed on the address bus by the microprocessor. This addresses another memory register and the contents of that register are read in and placed in an 8-bit register inside the microprocessor called the instruction register.

What action the microprocessor takes next is dependent entirely on the value of the 8-bit word in the instruction register. As the name of the register implies, this 8-bit word is treated as an instruction to the microprocessor defining its next actions. It is

158

worth noting that the designer of the microprocessor itself determined that the addresses FFFE and FFFF should be issued when $\overline{\text{RESET}}$ is asserted, while the designer of the microprocessor system determines what the contents of those memory locations are at the time of ROM programming. The system designer, in deciding the contents of locations FFFE and FFFF, has determined the next step in the process.

In specifying the contents of all other ROM locations, the system designer completely specifies the course of action which the microprocessor will follow, and makes use of what is termed the *deferred design* property of the microprocessor. This extremely important property of the device is, in essence, that the designer of the microprocessor itself need have no knowledge of the final use to which the device will be put. The decision on the actual usage is deferred until the system as a whole is designed.

deferred design

An example of a data read action

We will assume that what the designer wants the microprocessor to accomplish is to read an 8-bit data word from the A/D converter to an internal processor register. We will assume that data coming from the A/D converter to the PIA can be accessed by the microprocessor performing a *read* of register 0 of the PIA. The steps which the processor executes are:

(a) The 16-bit content of the microprocessor internal register called the program counter is placed on the address bus, and the contents of the memory location so specified are read in to the 8-bit instruction register. The system designer must have determined all the processor operations prior to this step, so that the 8-bit instruction word read from memory causes the following subsequent actions.

(b) The contents of the program counter are increased by one, the new contents put on the address bus, the contents of the new memory location read into the processor and that 8-bit word stored internally as the high-order byte of an address word.

(c) The contents of the program counter are increased by one again, the new contents put on the address bus, the contents of the memory location read into the processor and stored as the low-order byte of an address word.

(d) The address produced by combining these two 8-bit bytes is placed on the address bus and the R/W is set to *read*. The address must have a value which addresses register 0 of the PIA, for example (from Figure 122 the address 0400 (hex). This causes the contents of register 0 of the PIA (which is the A/D converter output word) to be read in to an internal register of the microprocessor called the accumulator, and so a *read* of the A/D converter output (representing the voltage level of one of the batteries under test) has been executed.

159

(e) Finally, the program counter contents are incremented by one again ready for the next action the microprocessor is required to execute.

As you can see from these steps, somewhere in memory there must have existed a sequence of three adjacent 8-bit words, and prior to these processor actions, the contents of the program counter must have been set to specify the first of the three memory addresses. Figure 123 illustrates the situation before and after the execution of the sequence, assuming that the three memory words are stored at hex addresses FF30, FF31 and FF32.

Figure 123

	program counter	memory address	contents	
before	FF30	FF30	B6	← instruction code (read)
		FF31	04	← high-order address byte
		FF32	00	← low-order address byte
after	FF33	FF33		← next instruction code

An example of a data store action

This time we will assume that the designer wants the data word stored in the accumulator by the previously described read action to be stored at address 0035 (hex) which Figure 122 indicates is a valid RAM address for the system configuration. The program counter contains FF33 (hex) as a result of the immediately preceding read action. The processor steps (with much shortened descriptions) are:

(a) Read the contents of address FF33 to the instruction register.

(b) Read the contents of address FF34 to the high-order byte of an address word.

(c) Read the contents of address FF35 to the low-order byte of an address word.

(d) Write the contents of the accumulator to the memory register specified by the address word.

(e) Increment the program counter contents to the value FF36.

Figure 124

	program counter	memory address	contents	
before read	FF30	FF30	B6	← instruction code (read)
		FF31	04	← high-order address byte
		FF32	00	← low-order address byte
		FF33	B7	← instruction code (write)
		FF34	00	← high-order address byte
		FF35	35	← low-order address byte
after store	FF36	FF36		← next instruction code

Figure 124 shows the full sequence of memory contents required to control both the read and write operations, and the contents of the program counter before the read and after the write. As a result of these two steps, location 0035 (hex) of the RAM would contain the digital representation of the voltage of one battery.

Some further general points

In this read and store example, each group of three memory address contents is called an *instruction*. Each of these instructions contained two bytes of address information, they were *memory reference instructions*. The MC6800 processor has been designed to recognize 72 different instructions, not all of which have an address associated with them, and many of which occupy only one or two memory locations. However, to proceed any further along that line of thought would take us outside the limits of this book.

instruction

A set of instructions which controls the microprocessor to perform a series of actions and hence to execute some predetermined useful task is called a *program*. The designer has the job of creating the program which causes the system to satisfy his performance criteria. The design, testing and de-bugging of a microprocessor system is very demanding and time consuming and usually forms the major component (typically in excess of 60 per cent) of the overall system design task. When the program is known to be correct, it is usually put into ROM or PROM. Keeping the program in ROM means firstly that it is always available at system switch-on, and secondly that it will not get overwritten by some malfunction of the system hardware or its program.

program

RAM is used in the system for the storage of data which is read into the system (such as the digital representations of the battery voltages) or generated within it (such as the successive multiplexer channel numbers which must be generated by the system) during the execution of the program. Such data cannot be stored in the ROM because its nature is unknown prior to operation of the system, or is required to change during operation.

The role of the PIA in the system

We have already said that the basic role of the PIA in this microprocessor system is to control the action of the multiplexed A/D converter and to allow its output to be read into the microprocessor. However, the device is a general-purpose component of microprocessor systems. *PIA* stands for *peripheral interface adaptor*, and it is one of several devices produced as part of the MC6800 'family of parts'. It is a device which is designed to allow the transfer of digital data between the data bus of the microcomputer and devices external to the system (*peripheral* devices). The PIA requires some description before we can consider its role in the

peripheral interface adaptor

specific application. The device is quite complex, a result of the desire on the part of the designer to create great flexibility of use (another example of deferred design). This description will be simplified and incomplete, we shall discuss only those features of the device which are relevant to the specific application.

Figure 125 shows the connections on the microprocessor side of the PIA. Eight bidirectional lines connect the data bus to bus conversion circuits which create two unidirectional buses inside the PIA. Address lines supply address information from the microprocessor to three select inputs (CS0, CS1 and $\overline{\text{CS2}}$) and two register select inputs RS0 and RS1. The control bus supplies the read/write signal, φ_2 and the $\overline{\text{RESET}}$ signal. These eight lines supply the chip select, register select and read/write control circuits in the PIA. Signals $\overline{\text{IRQA}}$ and $\overline{\text{IRQB}}$ are generated within the PIA and will be discussed shortly.

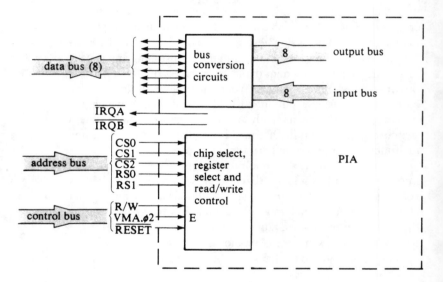

Figure 125

Within the PIA, the output and input buses both connect to six internal registers to enable the microprocessor to write data to or read data from each register. The PIA is functionally divided into two halves, the A half and the B half. Each half contains a control register, a data direction register and an output data register. Each of these registers can have data written into them by the microprocessor, and can have their contents read by the microprocessor. The contents of the control register determine the specific functions to be performed by the PIA. The contents of the data direction registers determine whether data will be passed through the PIA from microprocessor to peripheral device or through the PIA from the peripheral to the microprocessor. The output data registers

162

receive the data from the microprocessor which are to be sent to the peripheral device.

The peripheral side connections of the PIA are shown in Figure 126. The contents of the output data registers are fed to peripheral interface circuits, each of which can connect to an 8-bit peripheral bus. The interface circuits can act as sources of data on the peripheral bus lines, or as receivers of data from them depending on the data in the data direction registers. When the interfaces act as sources, it is the data contained in the output register which are put onto the peripheral bus. When the interfaces act as receivers of data, the incoming data are read from the peripheral bus.

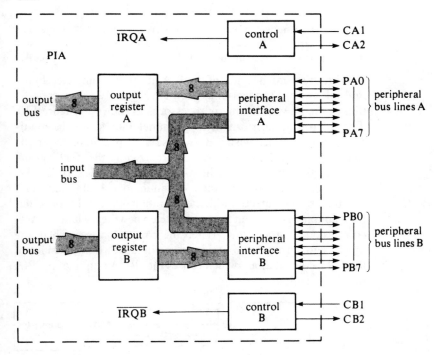

Figure 126

There are also four control signals on the peripheral side, two in each half of the PIA. These can be configured to act in a variety of ways depending on the data contained in the control registers. CA1 and CB1 can only receive signals, while CA2 and CB2 can be configured to receive signals or to send them. $\overline{\text{IRQA}}$ and $\overline{\text{IRQB}}$ are 'interrupt request' signals generated by the control signals CA1, CA2 and CB1, CB2. The content of the control registers determines which control signal generates the interrupt request and whether a 1–0 transition or a 0–1 transition results in the assertion of the interrupt request line. Their function will become a little clearer when we discuss the specific PIA application.

Without considering the detail of how it is achieved, we will assume that the microprocessor has written data words to the two control registers and to the data direction registers so as to set up the following PIA characteristics.

(i) Peripheral interface B is configured to drive the peripheral bus lines B, by putting the contents of output register B on them.

(ii) Peripheral interface A is configured to accept data from peripheral bus lines A.

(iii) CA2 is configured as an output control line which is asserted when the microprocessor performs a read operation of the data on peripheral bus lines A. It is cleared of the assertion when a 0−1 transition occurs on CA1.

(iv) A 0−1 transition on control line CA1 causes the IRQA line to be asserted. The assertion is removed when the microprocessor performs a read operation of the data on peripheral bus lines A.

With this configuration of the PIA, it can be connected to the multiplexed A/D converter as shown in Figure 127. The connection of the PIA to the microprocessor is as shown in Figure 120. The microprocessor can write a channel address for the multiplexer into output register B to cause one specific battery to be selected. It can then send the convert signal to the A/D converter by performing a read on the A-side input and so asserting CA2. The microprocessor can then continue to execute some task completely unconnected with the operation of the A/D converter, for example printing out data on the teletypewriter, while the A/D conversion is proceeding. When the conversion is finished, the 'done' signal from the A/D converter will cause a 0−1 transition of the CA1 control input, which causes the $\overline{\text{IRQA}}$ output of the PIA to be asserted.

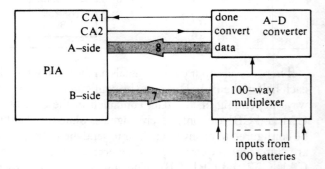

Figure 127

Figure 120 shows that $\overline{\text{IRQA}}$ of the PIA is connected to $\overline{\text{IRQ}}$ of the microprocessor (the 'interrupt request' input to it). We mentioned earlier that an assertion of this line could cause the microprocessor to temporarily stop executing its current task and perform

164

some different task. In this case, the assertion of the 'interrupt request' line causes the microprocessor to execute a read operation on the A-side data input (which is the A/D converter output). This completes the control of the A/D converter which started with the writing of a channel number into the PIA by the microprocessor. It can then either be controlled to issue the next channel number, and to proceed with its data printing until the next 'conversion complete' signal is received or to process the data it has just received.

The role of the ACIA in the system

In the previous section, the PIA was used to implement parallel data transmission between the A/D converter and the data bus of the microcomputer. In much the same way it would be possible to use a PIA as a means of transmitting 8-bit ASCII characters from the microprocessor to a parallel input teletypewriter, and as a means of receiving ASCII characters from the teletype. (The ASCII code for the digital representation of alphanumeric characters was discussed in Chapter 2).

However, because of the frequent requirement for a teletype to be a considerable distance from a computer it has become common practice to use *serial data transmission* to and from the teletype. This reduces the number of leads required to interconnect the teletype and the computer, which is important when long distances are involved.

serial data transmission

Serial data transmission

In serial data transmission, data words are sent one bit at a time, the whole data word being made up of a time sequence of 1s and 0s on a single line. Figure 128 shows the 8-bit word 01001011 (the ASCII code for the letter K) in serial form. It is conventional that the first (in time) bit of the word is the least significant bit, and the last bit is the most significant bit (the parity bit). Whenever a key on the teletypewriter is stuck, the digital representation of the character is transmitted serially from the teletype. To cause the teletype to print a character, a serial data word must be transmitted to it.

Serial data words to and from a teletype actually take the form shown in Figure 129. Because of the serial nature of the data, the receiver of the data must synchronize its operation with that of the transmitter. This means that the receiver needs a short time before receiving the first data bit to achieve this synchronization. The first information the receiver gets of the arrival of the word is a transition from the idle (high) state of the line to the low state. This first low state is called the start bit, and provides the time necessary for synchronization to be established. The eight data bits then follow the start bit, and finally either one or two stop bits (high logic states), the number used being fixed by the adoption of one or other convention by both devices. Figure 129 shows the ASCII code for the letter K in serial form using the two stop bit convention. What the stop bits really mean is that the idle state will last for at least two time intervals between successive character transmissions.

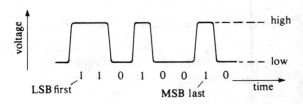

Figure 128

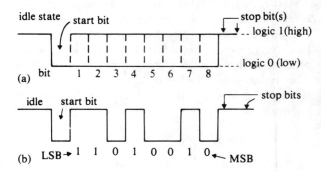

Figure 129

The time taken to transmit a serial character depends upon the basic clock rate used to generate the data in serial form. This clock rate is normally described by the maximum number of bits which can be transmitted each second. This is called the bits/second rate or *baud rate*. A common standard rate is 10 characters per second, with the serial format having two stop bits. Each character then contains 11 bits (1 start, 8 data and 2 stop) and the baud rate is $10 \times 11 = 110$. More recent teletypes run at 300 baud using only one stop bit (that is 30 characters per second) and even at baud rates of 9600 or 19200.

baud rate

The ACIA is a device which has been designed to convert data between the serial format and parallel data format. This means that it can be used' in microcomputer to convert between the parallel format of the microprocessor and the serial format required by such peripherals as Visual Display Units, printers and teletypes.

166

The asynchronous communications interface adaptor (ACIA)

Figure 130 shows a simplified block diagram of the device. This diagram does not show the internal connection between the control and status registers and the remainder of the components, nor does it show the chip select and read/write circuitry contained within the device. The transmit shift register is a circuit which converts parallel to serial data and the receive shift register converts from serial to parallel.

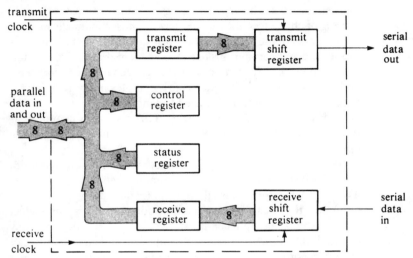

Figure 130

Eight-bit words on the data bus of a microcomputer can be written into the control register to set up the mode of operation of the device. The status register contains an 8-bit word which indicates the progress of a data word to, or from, the teletype on its way through the ACIA. The microprocessor can read the contents of this register to monitor the progress of a data transfer. A word to be transmitted from the microprocessor to the teletype printer can be written into the transmit register to be serialized. A serial word generated by the teletype is received by the receive shift register and transferred to the receive register. A read operation by the microprocessor can then collect the resulting 8-bit parallel word. The control register content determines the convention used by the ACIA in its serial transmission and reception of data. It determines the type of parity to be used in transmitted words, and the number of stop bits to be added. Similarly, it determines the type of parity to be checked in an incoming serial word, and the expected number of stop bits. It has other functions in connection with the generation of interrupts, and the transmission of data using telecommunication links.

167

The status register contents indicate the progress of a data word through the ACIA, and error conditions in the receipt of data. A parity error on a received word will set one of the status register bits, the absence of the correct number of stop bits sets another. One bit becomes set when a serial word has been received by the receiver shift register and passed to the receiver register. A fourth bit indicates that a word has been passed from the transmit register to the transmit shift register. A fifth indicates that more than one data word has been received by the ACIA without the first being read by the microprocessor, so indicating that a character (or characters) has been lost. During the transmission or receipt of a string of characters, the microprocessor must continually monitor the contents of the status register so as not to overwrite characters in the transmit register, or lose characters in the receive register. This implies that the communication between the microcomputer and the teletype is not simple, a fairly complex program of microprocessor instructions is required to control the process. With the aid of such a program, however, the ACIA does allow communication to take place between the microcomputer and a human operator.

As far as the microprocessor is concerned, communication with the ACIA is exactly the same as communication with memory. The ACIA has four internal registers (two read-only, two write-only) and these can be individually addressed using chip-select and register-selct inputs on the device. Figure 131 shows the way in which the ACIA is connected to the data and address buses of the microcomputer. The control signals read/write, φ_2, interrupt request, valid memory address, and so on, are together considered to form a control bus in the system.

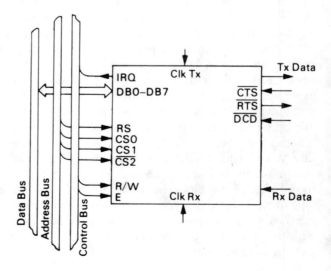

Figure 131

CS0, CS1 and $\overline{CS2}$ are chip-select inputs which will be driven by address lines to specify the address area occupied by the ACIA. The enable input E will be driven by the φ_2 clock to synchronize bus activity. RS (register select) and R/W (read/write) together specify which of the internal registers is being addressed. If RS is 1, the *read* specifies the status register, while *write* specifies the control register. If RS is 0, the *read* specifies the receiver register and *write* specifies the transmitter register. The teletypewriter cannot be connected directly to the serial-in and serial-out terminals of the ACIA. The ACIA output is not in the correct electrical format for receipt by the teletype. Its voltage levels, and the current drive available from the ACIA are not compatible with the teletype input circuits. Similarly, the voltage and current levels generated by the teletype are not compatible with the logic levels and current handlng capability of the ACIA receiver input. Another component must be placed between the ACIA and the teletype as shown in Figure 132 to convert voltage and current levels of the two signals. Its characteristics depend on the design of the particular teletype used. There are three common types of teletype output signal and input requirement. They are 20 mA current loop, 60 mA current loop and RS232C.

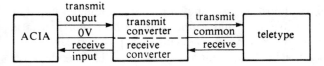

Figure 132

In a 20 mA current-loop system, the teletype sends a current of 20 mA around a 2-wire loop passing through the receive converter to represent a logic 1, and sends no current to represent logic 0. The converter must change these two possible current levels into the logic levels required by the ACIA. For signals to the teletype, the transmit converter must change the ACIA logic levels into the 20 mA and 0 mA currents expected by the teletype input.

A 60 mA current-loop teletype just uses 60 mA instead of 20 mA to represent a logic 1, and the converters must receive and generate these currents.

An RS232C teletype generates +12 V to represent logic 1, and −12 V to represent logic 0. The converters must change these logic level voltages to the +3.5 V and 0 V levels required by the ACIA, and vice versa.

The overall system functions

Having looked at the individual components of the microprocessor system which will act as a controller for the battery test system, we can now look at the way in which the overall control function is achieved. The program of instructions for the microprocessor contained in the ROM dictates each individual action which the microprocessor executes. It would not be useful to look in close detail at these small steps, but it is possible to see how sequences of instructions could achieve the desired control function.

The control of the A/D converter is accomplished as we have already seen, by a write operation of a channel number to the B side output register of the PIA, followed by a read operation of the A side of the PIA to obtain the A/D converter output data. A write into RAM can then store this data. A succession of 100 such sequences, each with a different multiplexer channel number, stores data on the output voltage of each battery. The next time the battery voltages are read, the new A/D converter output, instead of being stored, would be compared with the previously stored value. The comparison would be achieved by an instruction which causes the microprocessor to subtract one value from the other, and then subsequently to execute different sequences of instructions depending on whether the difference is zero or non-zero. If the difference is zero, then the battery voltage has changed by less than 1 part in 256 (eight bits of A/D converter output data provide a resolution of one part in 256) and the new reading can be ignored. If the difference is non-zero, the battery voltage has changed by almost 50 mV, and so the new reading is stored in place of the old one, and a print-out of the required data is initiated on the teletype. A character is sent to the teletype by a write operation in the transmit register of the ACIA. The character is transferred to the transmit shift register of the ACIA where it is serialized and sent to the teletype. The microprocessor reads the status register of the ACIA to determine when the transmit register is ready to receive the next character. The microprocessor's program of instructions has the task of converting a digital word into a sequence of ASCII numeric characters which, when printed, will be a denary representation of the voltage or of the multiplexer channel number identifying a specific battery. The microprocessor system did not include the method of generating the time data which was part of the required alpha-numeric print-out. The function of timekeeping and of generating digital data representing time is normally performed by a timer chip. This is another LSI device which connect to the address and data buses of the microcomputer. It is driven by a crystal-controlled oscillator and contains appropriate counting circuits to record the passage of time in hours, minutes and seconds. The time chip has address and enable inputs which are driven by the address bus, and the time information can be put onto the data bus to be read by the microprocessor whenever the time of day is required. This time information is also converted to ASCII characters by the microprocessor acting under the control of its program of instructions, and sent to the teletypewriter.

Summary:
A Microcomputer System

Having read this chapter, you should now be equipped to look at a diagram like Figure 133 on p. 172 (which is taken from the manufacturers literature on the MC6800) without feeling utterly confused. If you can, then we have achieved the major part of our original aim. We hope that you will also be able (but with less familiarity) to look at other microprocessor manufacturers' literature and obtain useful information from it.

We have not tried to describe the internal structure of any of the components of the microcomputer configuration of Figure 133. We have been concerned only with their functional characteristics, and then only to a limited extent. The full functional characteristics are sufficient to understand, and indeed design, microprocessor-based systems. With the collection of components of a system to which you have been introduced, and a familiarity with the means of interconnecting those components, you should be able to begin to understand any microprocessor-system configuration you might meet.

In this rather long chapter, we bagan with digital components which you had met in earlier chapters, simple gates and registers and developed the idea of a *simple memory* to hold several 8-bit data words. You met the idea of a *register address* and saw the need for *address decoding* to allow unique selection of a single register. We explained how this address decoding could be achieved using a ROM, or by the use of logic gates.

In considering the reading of the data from the simple memory to some data receiver, we introduced *open-collector gates* and *tri-state outputs*, which allow a common set of data lines to carry information from one of several registers, provided only one register is selected at any time. This led on to the idea of a *bus*, and in particular of separate *address* and *data buses*.

We introduced the practical implementation of the simple memory, the *random access memory (RAM)*. All the data registers, the address decoding, the tri-state output buffers are included within the device. Because the RAM has only one set of connections for data being written into memory and data being read from it, it has to be connected to a *bi-directional data bus*. A practical memory device also has *chip select* facilities, by means of which the

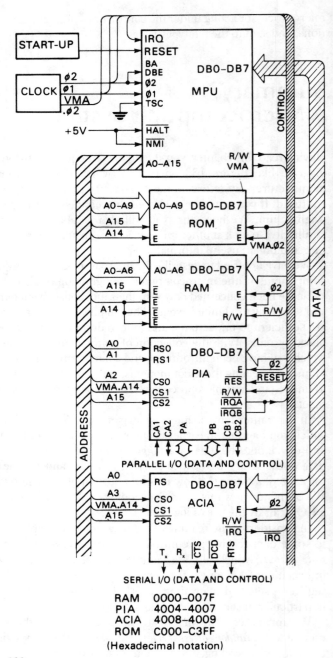

RAM	0000–007F
PIA	4004–4007
ACIA	4008–4009
ROM	C000–C3FF

(Hexadecimal notation)

Figure 133

172

device can be *enabled* to perform its designed function, or *disabled* and effectively disconnected from the data bus. A RAM also needs to be supplied with a *read/write signal* to indicate whether it is required to store data or send it out.

We then went on to show how memory devices could be combined together to produce large memories and how the chip select facility enables each device to be allocated a specific place in the whole memory space. *Memory maps* were used to indicate which device contains which addresses. When large numbers of devices are involved, the chip select facility does not provide sufficient addressing flexibility, and *address decoders* have to be used. Such address decoders can be implemented using a ROM, or special-purpose MSI (medium scale integration) devices such as the *3-to-8 decoder*. You also met *timing diagrams* used to indicate the timing requirements of memory elements. (ROMs) are devices used to provide *non-volatile memory* whose data are preserved during switch-off. We described a *mask-programmed ROM*, a *field programmable FPROM* and an *erasable programmable EPROM*. Which of these devices is used in a final product is an economic decision based mainly on the quantity to be produced, but each of the devices has a specific role to play in the initial development of a system.

The *microprocessor* was introduced as a versatile receiver and source of digital data. We described some basic features of the MC6800 microprocessor as an example of a wide range of similar devices. You met the signals required by the processor, and those generated by it, and learned that without memory, a microprocessor cannot perform any useful function.

A specific combination of microprocessor, memory and other devices together make up a *microcomputer*. We showed how a memory map can be created for such a system, and how the form of the memory map depends on the way in which address lines are used by the devices. We also described the action taken by the microprocessor in certain conditions, so as to illustrate the way in which data are moved around the system, and in particular to show the sequential step-by-step nature of a microprocessor system. You met the idea of *computer instruction*, of a *program* as a sequence of instructions, and the *deferred design* property of the microprocessor which makes its use applicable in so many different situations.

The PIA is used to control the A/D converter by permitting the communication of channel number and A/D converter output between the microprocessor and the converter, by generating the 'convert' signal and by receiving the 'conversion done' signal. The PIA itself is designed to be very flexible in the use to which it is put. Data must be written into its control and data direction registers by the microprocessor to establish a specific configuration of the device.

The ACIA is used to convert parallel digital data inside the microcomputer to the serial format required by the teletypewriter, and vice versa.

By writing data to the control register, and reading data contained in the status register of the ACIA, the microprocessor controls the progress of data tranfers to and from the teletypewriter. An additional component is required between the ACIA and the teletype to establish the required voltage and/or current requirements of each device. This component is called a *converter*, and is normally a 20 mA current loop, a 60 mA current loop or an RS232 converter. The complete controller function is established by the program of instructions which exists in the ROM.

Problems for Chapter 5

1. If a data receiver was required to receive the contents of any one of sixty-four 8-bit buffer registers, how many address bits would be needed to uniquely specify one register? How many tri-state buffers would be needed?

2. If, in the arrangment of Figure 111 the least significant address bit is used to drive $\overline{CS}$ instead of the most significant bit, sketch the resulting memory map.

3. Using the scheme shown in Figure 115, show how two extra address bits can be used to enable or disable three 3-to-8 address decoders, so as to enable the memory size to be made three-times as big (that is, 24 K × 8-bit memory) without using any inverters on the additional address lines. The memory size can be increased to 32 K × 8-bit using the same number of address lines and one inverter. How?

4. Using the diagram shown in Figure 120 determine for each of the following hexadecimal addresses, whether the microprocessor is addressing a register in RAM or ROM or neither.

   (a) FFF   (b) EO32   (c) DE85   (d) AOOF
   (e) 8A1B   (f) 4444   (g) 3FF3   (h) 10A1
   (i) OOOO

5. The following questions all relate to the microcomputer configuration of Figure 133.

(a) How may ROM registers are there?

(b) How may RAM registers are there?

(c) For each of the following hexadecimal addresses, decide whether the address refers to the ROM, the RAM, the PIA or the ACIA or none of these.
FFFE, D300, 937A, 7A04, 5BB8, 500C, 2222, 003F.

(d) Construct a full memory map of the microcomputer system.

Appendix A

Many of the examples and ideas described in this book draw upon the basic rules of network theory. Although it was not our intention to teach these rules, we thought it might be useful to provide a convenient reference for the most important ones.

A1   *Ohms relationship* states that the resistance of a circuit or component is defined as the ratio of the voltage across it to the current flowing in it, that is

$$R = \frac{V}{I},$$

Which can also be stated as $V = I.R$.

A2   *Kirchhoff's current law* states that the algebraic sum of the currents at a node is zero. This means that the total current flowing into a node is equal to the total current flowing out of the node.

A3   *Kirchhoff's voltage law* states that the algebraic sum of the voltages across all the components around any closed loop in a circuit is zero.

A4   *Series combination of resistors.* The equivalent resistance of resistors $R_1, R_2, R_3 \ldots R_n$ connected in series is equal to the sum of the resistors $R_1 + R_2 + R_3 \ldots + R_n$.

A5   *Parallel combination of resistors.* The equivalent resistance of resistors $R_1, R_2, R_3 \ldots R_n$ is calculated from the expression.

$$\frac{1}{R_{equivalent}} = \frac{1}{R_1} + \frac{1}{R_2} + \frac{1}{R_3} + \frac{1}{R_n}.$$

A6   The *voltage divider rule* states that the voltage across two resistors in series divides between them in the ratio of the values of their resistance. The larger resistance has the larger voltage drop.
The voltages V_1 and V_2 of Figure A.1 are given by

$$V_1 = V \frac{R_1}{R_1 + R_2}$$

$$V_2 = V \frac{R_2}{R_1 + R_2}.$$

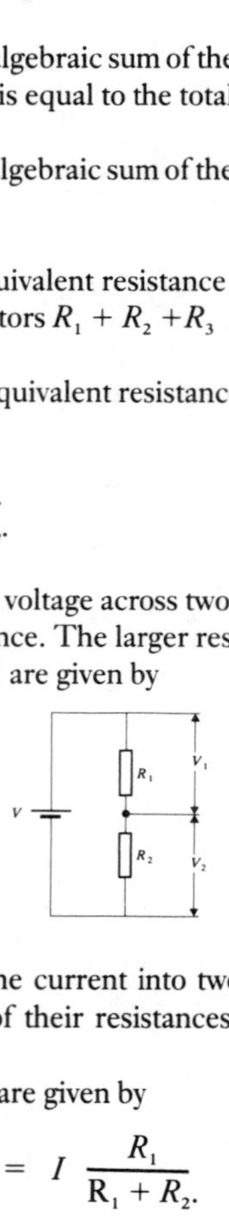

Figure A.1   The Voltage Divider Rule

A7   The *current divider rule* states that the current into two resistive components in parallel divides between them in the inverse ratio of their resistances. The smaller resistance takes the larger current.
The currents I_1 and I_2 of Figure A.2 are given by

$$I_1 = I \frac{R_2}{R_1 + R_2} \qquad\qquad I_2 = I \frac{R_1}{R_1 + R_2}.$$

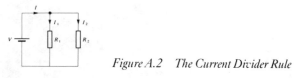

Figure A.2   The Current Divider Rule

176

A8 *Step response of an RC network.* When the input voltage to the RC network of Figure A.3 changes from 0 to V, the output voltage is given by the expression

$$v_0 = V(1 - e^{-t/RC}).$$

RC is called the time constant of the circuit.

The output voltage measured as a function of time is shown in Figure A.4. When $t = RC$ the

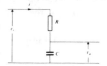

Figure A.3   An RC network

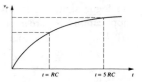

Figure A.4   Step Response of an RC network

output voltage equals 63 per cent of the input voltage, and when $t = 5RC$ the output equals 99.3 per cent of the input.

A9 An ideal *voltage source* is defined as a voltage generator whose voltage output is independent of the current delivered by the generator.

A10 An ideal *current source* is defined as a current generator whose current output is independent of the voltage between the two terminals of the generator.

A11 The *voltage gain* of an amplifier is defined as the ratio of the output voltage to the input voltage

$$\text{Gain (volts)} = \frac{V_{\text{out}}}{V_{\text{in}}}.$$

The gain may also be expressed in decibels as $20 \log_{10} (V_{\text{out}}/V_{\text{in}})$.

Appendix B

Operational amplifiers

An operational amplifier is an integrated circuit (i.c.) device used to perform arithmetical operations on analogue signals. Although the basic theory of operation should be familiar to you, there are four applications that are important enought to warrant revision in this appendix.

(*i*) *Unity gain buffer.* Figure B.1 shows a block diagram for a differential input operational amplifier. The

Figure B.1   Differential Input Operational Amplifier

input voltage V_i is connected to the non-inverting input. The output voltage is V_o and the output is connected to the inverting input. The open circuit gain of the amplifier is A_v.

The output of the amplifier is given by

$$V_o = A_v (V_+ - V_-),$$

that is the open circuit gain of the amplifier times the difference in the voltage at the input terminals. Now V_+ is equal to V_i the input voltage and V_- is equal to V_o because of the direct connection. We can therefore write the output voltage as

$$V_o = A_v (V_i - V_o)$$

which can be rewritten as

$$(1 + A_v) V_o = A_v V_i,$$

so that

$$V_o = \frac{A_v}{1 + A_v} V_i .$$

for most operational amplifiers A_v is very large (>1000) so we can ignore the 1 in the denominator, hence

$$V_o = V_i,$$

and the amplifier has unity gain. Because the output is connected to the inverting input, this is an example of an amplifier with *negative feedback.*

The input impedance of the amplifier with feedback is

$$Z_i A_v,$$

where Z_i is the input impedance of the differential amplifier, without feedback, times the open circuit gain.

The output impedance of the amplifier with feedback is

$$Z_o / A_v,$$

the output impedance of the open circuit amplifier divided by the open circuit gain.

The high input impedance obtained with feedback ensures that the amplifier does not load the input signal source and the low output impedance enables the amplifier to drive long cables. The amplifier behaves as an impedance converter.

(*ii*) *Fixed gain amplifier.* The operational amplifier can be used to provide a wide range of gain. Consider the circuit shown in Figure B.2. Current I_i flows through Z_i to junction S, and current I_F flows through Z_F the feedback resistance. Feedback ensures that the input impedance of the amplifier will be high, so that current I_- which flows into the inverting input will be small compared to I_i and I_F. Since the sum of

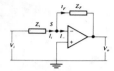

Figure B.2   Fixed Gain Amplifier

the currents flowing into and out of the junction S must be zero (Kirchhoff's rule) then

$$I_i = I_F + I_-.$$

If I_- is small then

$$I_i \simeq I_F.$$

The non-inverting input is at 0 V, hence $V_+ = 0$. The current I_i is equal to the voltage drop across Z_i divided by Z_i, hence

$$I_i = \frac{V_i - V}{Z_i}.$$

Likewise the current I_F equals the voltage drop across Z_F divided by Z_F, so

$$I_F = \frac{V - V_o}{Z_F}$$

but $I_i \simeq I_F$, so that

$$\frac{V_i - V}{Z_i} = \frac{V - V_o}{Z_F}. \tag{B.1}$$

$V_o = A_V \cdot V$ where $V = (V_+ - V_-)$, but since $V_+ = 0$ we may write V_o as

$$V_o = -A_v V_-,$$

but $V = V_-$ so that

$$V = -V_o/A_v.$$

Substituting this into equation (B.1) then

$$\frac{V_i + V_o/A_v}{Z_i} = \frac{-V_o/A_v - V_o}{Z_F}$$

Collecting and rearranging terms:

$$\frac{V_i}{Z_i} = \frac{-V_o}{Z_i A_v} \qquad \frac{-V_o}{Z_F A_v} \qquad \frac{-V_o}{Z_F,}$$

therefore

$$\frac{Z_F}{Z_i} V_i = -V_o \left[\frac{Z_F}{Z_i} \frac{1}{A_v} + \frac{1}{A_v} + 1 \right].$$

If A_v is very large, the first two terms in the square brackets are approximately equal to 0, hence

$$\frac{Z_F}{Z_i} V_i = -V_o.$$

The gain of the amplifier is defined as the ratio V_o/V_i so rearranging

$$\frac{V_o}{V_i} = \frac{-Z_F}{Z_i}.$$

This is a very simple and useful result which can be obtained in a much quicker way as follows. If A_v is very large, then, since $V = -V_o/A_v$, V must be very small compared to V_o. The inverting input can be thought of as being at zero potential (0 V). Thus the point S, though not directly connected to 0 V is virtually at zero volts. It is called a virtual earth point.

Putting $V = 0$ in equation (B.1) we have

$$\frac{V_i}{Z_i} = \frac{-V_o}{Z_F}$$

and if follows directly that

$$\frac{V_o}{V_i} \simeq \frac{-Z_F}{Z_i}$$

If $V \simeq 0$ then I_- must also be effectively zero, which implies a high input impedance at the inverting input.

Figure B.2 shows the non-inverting input connected to the common line, which is perfectly acceptable for an ideal op-amp.

However, in the case of a real amplifier it is normal to insert a resistor between these two points to correct for small d.c. offsets caused by differences in the inverting and non-inverting inputs.

An approximate value for this resistor is the equivalent resistance of the parallel combination of Z_i and Z_F. Figure B.3 shows a circuit diagram for a fixed gain amplifier, including the power supplies, V_+ and V_-.

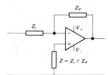

Figure B.3 *Complete Circuit for Fixed Gain Amplifier*

(*iii*) *The summing amplifier (inverting adder).* Occasionally we need to be able to add voltages from several sources to produce a single output voltage. This can be achieved using a circuit similar to that shown in Figure B.4. The three input voltages V_1, V_2 and V_3 are connected to the virtual earth point of an op-amp via the three resistors R_1, R_2 and R_3 respectively.

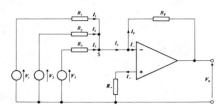

Figure B.4 *The Summing Amplifier*

The total current entering into junction S is $I_1 + I_2 + I_3$, but $I_1 = V_1/R_1, I_2 = V_2/R_2$ and $I_3 = V_3/R_3$ since S is the virtual earth point.

The current through R_F must equal the sum of the currents flowing into S, because $I_- \simeq 0$, therefore

$$I_F = \frac{V_1}{R_1} + \frac{V_2}{R_2} + \frac{V_3}{R_3} .$$

However, I_F is also equal to $-V_o/R_F$ so that

$$\frac{V_o}{R_F} = -\left(\frac{V_1}{R_1} + \frac{V_2}{R_2} + \frac{V_3}{R_3}\right) .$$

For the special case $R_F = R_1 = R_2 = R_3$, then the output voltage is given by

$$V_o = -\,(V_1 + V_2 + V_3) .$$

180

that is the sum of the input voltages times -1, hence the name inverting adder.

This circuit can be used to add voltages or combine them in various proportions. For example if $R_1 = \frac{1}{2}R_F$, $R_2 = 1/3R_F$ and $R_3 = 1/4R_F$, V_o will be equal to $-(2V_1 + 3V_2 + 4V_3)$. More voltages can be added by connecting them to S via suitable chosen resistors. Because the currents add at the virtual earth, this point is sometimes called the *summing junction*.

(*iv*) *The comparator*. The comparator is an example of a non-linear application of an operational amplifier. The basic function of the circuit is to produce an output voltage V_o which is related to the difference between an input signal and a reference voltage.

Most differential input operational amplifiers can be used as a comparator, if connected open-loop (that is no feedback) as shown in Figure B.5. The input signal is connected to the inverting input, and the reference voltage to the non-inverting input.

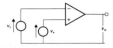

Figure B.5   Operational Amplifier as a Comparator

The output voltage $v_o = A(v_r - v_i)$, that is the open loop gain times the difference voltage. If $v_r > v_i$ then v_o is positive, whereas if $v_i < v_r$ the output is negative.

The maximum values of v_o are limited by the voltages supplied to the amplifier, but in some cases it may be more convenient to limit the maximum voltage swing, so that the comparator output can be connected directly to digital circuit components.

Special purpose comparators are available for this type of application.

Index

ACIA 161
AND gate 25
ASCII 21
Address 15, 124
Address decoder 124
Address decoding 139
Aliasing 109
Anti-aliasing filter 109
Analogue to digital converter 101
Aperture time 112
Asynchronous communications interface adapter 167

BCD code 18
BCD to 7 segment decoder 38
Binary arithmetic 31
Binary coded decimal 18
Binary counter 57
Binary digits 4
Binary word 15
Bipolar converters 96
Bipolar offset error 98
Bistable 58
Bit 15
Black boxes 2
Block diagrams 1
Boolean algebra 6
Buffer 24
Bus 16, 127
Byte 15

Carry 78
Carry output 78
Cascade 79
Clear direct 60
Clear state 58
Clocked register 60
Code 15
Code conversion 32
Conditional sequencer 70
Conversion rate 102
Conversion time 102
Counter Ramp Converter 101

D-type flip-flop 59
Data 15
Data lines 134
Decade counter 67, 79
Decoder (3 to 8) 139
Deferred design 159
Device address 134
Differential input multiplexer 115
Digital signal 4
Digital to analogue converter 88
Divide-by-*n* counters 77
Droop rate 111
Dynamic RAM 142
Dynamic range 100

EPROM 39, 148
End of conversion signal 102
Eraseable programmable read only memory 148
Even parity 21

FPROM 147
Feed through 113
Field programmable read only memory 147
Flip-flop 58
Full adder 32
Fusible PROM 148

Gates 13, 14

Half adder 31
Hexadecimal 20
Hold time 62

Instruction 161
Interface 87
Internal state 66
Internal state register 69
Inversion 14
Inverter 14, 24

J-K flip-flop 63

LSB 18
LSI 14
Latch 60
Logic function AND gate 14
Logic state 11

MOS 35
MSB 18
MSI 14
Mask programmed ROM 38
Masked programmed ROM 147
Memory device 58
Memory element 60
Memory register 60
Minterms 30
Modulo-10 counter 67
Modulo-n counter 67
Multiplexers 113

NAND 26
NOR 26
Natural binary 17, 18
Natural binary counter 57
Next state 62
Next state ROM 66
Number representation 17

OR gate 25
Octal 20
Odd parity 21
Offset binary 96
Open collector 130
Open collector NAND buffer 130
Output ROM 66

PIA 167
PROM programmer 148
Parity bit 21
Peripheral interface adapter 161
Positive logic 11
Present state 62
Preset counter 75
Preset number 74
Program 161
Programmable ROM 38
Programmable counter 81, 82
Propagation delay 46

Quantization 98
Quantization error 99
Quantization interval 98
Quantization levels 98
Quantization noise 99

$R-2R$ ladder network 94
RAM 41, 131
Random Access Memory 132
Read only memory (ROM) 13, 146
Reversible counter 75

SSI 14
Sample and Hold 110
Sampling rate 107
Sampling rule 107
Sequencers 68
Serial data transmission 165
Set direct 60

Set state 58
Set-up-time 62
Settling time 89
Seven segment display 32
Single ended multiplexer 115
States 8
Static RAM 142
Successive Approximation Converter 104
Sum of minterms 30
Synchronization 70
Synchronization (sequential circuits 70
System clock 151
System memory map 156

TTL 34
Timing diagram 142
Transducer 11
Transfer function 3
Tri-state buffer 129
Tri-state device 128
Truth table 7

Unipolar converters 96
Up-down counter 74, 75

VLSI 14
Venn diagrams 52
Volatile memory 146

XNOR 27
XOR 27